# VIRUSES
# AND
# REPRODUCTION

**Recent Titles in**
**Bibliographies and Indexes in Medical Studies**

Federal Information Sources in Health and Medicine: A Selected Annotated Bibliography
*Mary Glen Chitty, compiler, with the assistance of Natalie Schatz*

# VIRUSES AND REPRODUCTION

## A BIBLIOGRAPHY

Compiled by
ERNEST L. ABEL

BIBLIOGRAPHIES AND INDEXES IN MEDICAL STUDIES, NUMBER 2

GREENWOOD PRESS
NEW YORK • WESTPORT, CONNECTICUT • LONDON

**Library of Congress Cataloging-in-Publication Data**

Abel, Ernest L., 1943-
Viruses and reproduction : a bibliography / compiled by Ernest L. Abel.
p. cm. — (Bibliographies and indexes in medical studies, ISSN 0896-6591 : no. 2)
Includes bibliographies and index.
ISBN 0-313-26439-2 (lib. bdg. : alk. paper)
1. Virus diseases in pregnancy—Bibliography. I. Title. II. Series.
[DNLM: 1. Bacterial Infections—bibliography. 2. Pregnancy Complications, Infectious—bibliography. 3. Reproduction—bibliography. 4. Virus diseases—bibliography. ZWQ 256 A139v]
Z6671.2.V5A24 1988
[RG580.V5]
016.6183—dc19
DNLM/DLC
for Library of Congress 88-16575

British Library Cataloguing in Publication Data is available.



Library of Congress Catalog Card Number: 88-16575
ISBN: 0-313-26439-2
ISSN: 0896-6591

First published in 1988

Greenwood Press, Inc.
88 Post Road West, Westport, Connecticut 06881

Printed in the United States of America

The paper used in this book complies with the Permanent Paper Standard issued by the National Information Standards Organization (Z39.48-1984).

10 9 8 7 6 5 4 3 2 1

# Contents

# Preface

This bibliography contains references relating to the effects of viruses on reproduction, especially pregnancy and fetal development. The terminal date for inclusion in the present bibliography is 1986.

Although my aim was to make this bibliography exhaustive, many relevant entries may have been omitted. Most entries were taken from my personal collection of materials. Entries were also collected from the reference lists included in these publications. I also made use of many data bases now available, e.g., Medline, but more often than not, these data bases lagged considerably in their entries and were much more limited in focus.

Entries are arranged alphabetically by author. All items are numbered consecutively and are referred to by number in the Subject Index.

In compiling a volume such as this, the services and cooperation of a great number of people are to be gratefully acknowledged. Among those who deserve special thanks are Caryl Scheuer, Jean Brennan, Debbie Paduchowski and Judy Wegener.

# Introduction

Viruses are one of the three main causes of infectious disease, the other two being bacteria and fungi. Of the three, viruses are by far the most common source of acute disease and they are also a major contributor to chronic disease. More than 400 different viruses causing problems ranging from mild respiratory distress to cancer are known to affect humans.

Although this bibliography deals mainly with the consequences of viral infections, entries for some bacterial infections such as toxoplasmosis and syphilis have also been included because they are part of the TORCH categorization of congenital infectious diseases (see below).

Viruses are the smallest units of life capable of self-replicating. Unlike bacteria, viruses do not divide by binary fission. Instead, viral components are produced inside the infected cell and are assembled there before the virus is released.

Although there are a number of different definitions of viruses, most definitions emphasize four properties: 1) Viruses are entities composed of nucleic acids and protein. These nucleic acids are either deoxyribonucleic acid (DNA) or ribonucleic acid (RNA), never both, and are enclosed in a protein shell called a capsid. 2) Viruses replicate inside living cells, 3) Viruses redirect the host cell's machinery to synthesize materials for their own needs, and 4) Viruses affect the host cell causing either cell death, uncontrolled cell growth, or cellular malfunction. Classification of viruses is based upon the kind of nucleic acid they contain, e.g., DNA or RNA, the structure of their capsid, and the presence or absence of a lipid envelope.

The first step in viral infection begins with the random adsorption of a virus to the surface of a susceptible cell. Once contact is made, there is an interaction between specific sites on the surface of the virus and specific sites on the surface of the cell. It is this specificity of reactive and receptive site that is responsible for the selectivity of certain viruses. For example, poliovirus only adsorbs to human and other primate cells and only to specific types of cells like those in the brain. Receptors may also be present only during certain stages of development which is why some viruses pose a danger only for the newborn. Coxsackie B virus adsorbs only to receptors in newborn mice; it does not adsorb to adult cells. Other viruses have a specificity which is determined at a later stage of interaction.

After the virus has interacted with a receptor, the next step is penetration by the viral genome into the cells' cytoplasm. For some viruses with an enveloped capsid, only the part of the virus containing the genome enters; for those without an enveloping capsid, the entire virus enters.

As the virus enters the cell it goes into a phase called "uncoating" during which the nucleic acids are released from their capsids. The virus then takes over the cell's genetic material by a number of different processes depending on the nature of the virus. If it is a virus whose genome is made of RNA, it may act as a messenger RNA (mRNA) and initiate biosynthesis of its own protein. Retroviruses, on the other hand, have their RNA converted into DNA which is then incorporated into the cell's DNA where it then serves as a template for mRNA synthesis. DNA viruses (e.g., herpes) synthesize their own mRNA and use the host cell's enzymes to make more DNA and protein coats.

After viral nucleic acids and protein have been produced, the viral materials are assembled into new mature viruses. The final stage involves the release of these new viruses from the cell. Some viruses achieve their release as a result of the destruction of the cell which accompanies the take over of the cell's DNA. Others are released from the cell without the cell being destroyed.

The clinical effects of a viral infection are related to the host's response and its defense mechanisms. During infection, most viruses enter the blood through lymph channels. There is then a maturation period when the primary viremia spreads through the body and accumulates in particular target areas like the central nervous system (CNS) where it establishes localized infections. This site=specific accumulation is due to the receptor specificity of these cells. While in this incubation phase, the immune system attempts to destroy the virus.

The kind of cellular damage a virus produces varies with the virus itself. By taking over the cell's protein=manufacturing capacity, the virus shuts down the cell's ability to reproduce and repair itself. This may result in the breakdown of the cell membrane which begins to leak allowing extracellular fluid to enter the cell causing it to swell. Lysosomal enzymes may then leak from lysosomes within the cell resulting in further destruction and cell death.

Another kind of damage is fusion of infected cells. This results in the sharing of a common cytoplasm and creation of giant cells. A third kind of damage is increased cell replication resulting in neoplasms, e.g., warts caused by papilloma virus.

A number of viruses which cause only mild reactions in the adult pose a special hazard to the fetus and newborn. Susceptibility is in large part due to immaturity of fetal and newborn defense mechanisms. Major damage occurs if the infection occurs during the critical period of prenatal development when organs are forming.

In cases of in utero infection the virus passes from the mother to the fetus and the infection is said to be vertically transmitted. Perinatal infections are transmitted shortly before, during, or shortly after birth by contact with the virus during passage through the vagina.

## TORCH

The acronym TORCH is used to refer to five congenital infections which have many similar characteristics and pose the greatest threat to the fetus. These are toxoplasmosis (T), other agents, like syphilis (O), rubella (R), cytomegalovirus (C) and herpes simplex (H).

<u>Toxoplasmosis</u> is a bacterial rather than a viral infection and produces effects similar to cytomegalovirus infection. The incidence of infection among animals such as cats and dogs, sheep, pigs and cattle is as high as 60%. Infection in adults comes from eating undercooked meats which contain the infecting bacteria or from handling pet feces. In the adult, symptoms of

infection include fever and fatigue which last about a month. Infection of the fetus occurs via transplacental passage. Focal infection occurs in the liver, spleen, eye and brain. In some cases symptoms in the newborn are so mild that he appears normal and only later manifests signs of infection. The most serious consequence of infection is spontaneous abortion or stillbirth. General signs of infection include jaundice, a rash, anemia, fever, pneumonia, and an enlarged spleen. When the infant is several months old, he often appears lethargic and may convulse. About 25% of infected infants eventually suffer from neurological damage such as blindness and encephalitis. X-rays of brains from infected neonates are often characterized by diffuse calcification. Diagnosis of infection relies on detection of increased protein levels in cerebrospinal fluid and by observation of the bacteria in blood.

Other. This category includes a wide variety of infections, only the most common of which will be specifically mentioned.

(a) Syphilis is a bacterial infection transmitted almost exclusively by sexual intercourse in adults. It is associated with a number of well recognized symptoms in the adult including the appearance of a lesion which develops into a chancre. This disappears for a time and then numerous lesions appear. If untreated, most of those infected enter a long stage during which they have no symptoms of the disease. About one-third, however, experience destructive inflammation of various membranes including those in the central nervous system.

The fetus is infected transplacentally. Although the fetus can be exposed during its first six months of gestation, infection during this time does not appear to produce any congenital effects. Infection during the last trimester, however, can have serious consequences. About half the infants whose mothers are in the primary stage of untreated syphilis at this time are aborted, stillborn, or die shortly after birth. Of the other half, 50% may be normal and 50% may exhibit signs of infection. Symptoms of infection are similar to adult clinical symptoms except that chancres do not appear. Skin lesions are visible at birth, usually on the hands and feet. Anemia, enlarged liver and spleen, and jaundice are common. The lymph nodes are considerably enlarged. Bone damage is common. Another set of common physical defects is saddle nose and deformities of the upper central incisors. If treatment with penicillin occurs soon after birth, permanent damage can be avoided.

(b) Enteroviruses are nonenveloped RNA viruses which are often divided into three categories, polioviruses, echo viruses, and Coxsackie viruses. These viruses are most active during summer and fall and periodic outbreaks occur about every 5 years. Except for polio infection, symptoms of enterovirus infection in adults are usually mild although in some cases symptoms include heart disease. These viruses mainly multiply in the intestinal tract and cause "stomach flu" and vomiting.

Polio infection produces a wide range of symptoms from mild reactions to infantile paralysis. Most polio infections are so mild they go undetected and include headache, nausea, vomiting, sore throat and muscle soreness. More severe symptoms include encephalitis and meningitis that are signalled by stiffness and soreness in the neck, headache and drowsiness. In most cases, symptoms disappear but in a small percentage of individuals, paralysis eventually occurs, affecting mainly the legs, although other limbs can also be affected. Treatment involves complete bed rest. Prevention is now possible with vaccination. Prior to immunization against the poliovirus, infection generally resulted in high rates of postnatal mortality or paralysis in survivors.

Adult reactions to Coxsackie and echo viruses are fever, sore throat, and muscle soreness. Coxsackie infection also causes "hand, foot and mouth" disease, a condition associated with a rash on the palms of the hands, the soles of the feet, and in the mouth. Additional symptoms are sore throat and abdominal pain. Following incubation in the intestine, the Coxsackie and enterovirus may spread to the brain causing nervous system damage including meningitis and encephalitis. If infection occurs during the perinatal period, different effects occur depending on the nature of infection.

Infection by the Coxsackie virus increases the risk of heart disease including congestive heart failure, arrhythmia, increased heart size, and tachypnea. Infection can also occur in the liver, pancreas and CNS. There is no current method for preventing or treating either Coxsackie or echo virus infections.

(c) Chickenpox (varicella) generally attacks only cells in the skin and mucous membranes.

Maternal infection prior to the third trimester does not usually produce congenital anomalies. Infection during the perinatal period can result in transmission to the fetus/newborn. The incubation period is about 2 weeks. Symptoms of infection in the newborn range from a mild, pimply rash to pneumonia and encephalitis. Since most women have antibodies to chickenpox by the time of childbirth, infection is relatively uncommon among newborns.

The chickenpox virus is harbored in peripheral nerve cells. Remergence of the virus during adulthood causes skin eruptions and neuralgia, a condition called shingles (herpes zoster). The infection often emerges in the wake of emotional distress and illness. Symptoms include localized itching, pain and tenderness on one side of the body followed by the appearance of a rash, sores, then scabs which disappear and then reappear in rows. Effects on the fetus are similar to those seen in connection with chickenpox.

(d) Arboviruses are viruses carried by insects some of which produce encephalitis and other serious reactions. There are about 200 different arboviruses but only about 20 pose serious danger to the human fetus. California, Equine and St. Louis encephalitis are mosquito= borne. Symptoms are generally mild and consist of headaches and fever. However, severe reactions sometimes occur including kidney damage, and death. Colorado tick and Rift Valley fever, are tick-borne. Infection occurs mainly in spring and early summer and are primarily confined to wooded areas. Symptoms begin about 5 days after infection and include fever, chills, headaches and nausea in adults and in some cases meningitis.

Dengue Fever, a tropical virus infection carried by mosquitoes, is characterized by fever, aching joints and a rash in adults. Young children may experience excessive and often fatal bleeding. Lasa Fever, a fatal infection primarily confined to East Africa, is associated with bleeding of the internal organs and kidney damage.

Yellow fever is the most serious type of arbovirus infection. Symptoms include serious liver damage, jaundice and internal bleeding which often results in death.

(e) Hepatitis is an inflammation of the liver. Hepatitis can be caused by nonviral factors such as excessive consumption of alcohol. The viral cause is due to infection by two types of viruses. Hepatitis "A" infection is characterized in the adult by apathy, loss of appetite, nausea, vomiting, and fever. These symptoms begin about five weeks after infection. After the appearance of these initial symptoms there is a short period of improvement followed by the onset of jaundice, reappearance of fever, gastrointestinal distress and tenderness in the liver. Hepatitis "B" has the same symptomatology but occurs at a slower rate and the fever is higher. Hepatitis "A" is spread by ingestion of contaminated food or water. Hepatitis "B" or serum hepatitis, is spread by blood transfusion or use of contaminated needles. The latter is the more common type of infection but both types are widely distributed. Although a vaccine is not yet available, injection with gamma globulin soon after infection, will prevent symptoms.

(f) Mumps typically causes a swelling of the salivary glands, especially the parotid glands located just below the ear in the upper jaw. Infection during childhood results in discomfort in swallowing and rarely has any lasting effects. Adult infection can result in more diffuse effects including swelling of breasts and ovaries in women and damage to the testes in men which can lead to sterility. In 10% of all cases infection spreads to the brain causing muscular weakness, mental confusion, and paralysis in some cases.

(g) Slow virus infections are those which remain in the body for long times, often for life, and produce recurrent episodes of attacks which are progressively more serious and often result in death. Two of the most serious slow infections are Kuru and Creutzfeldt-Jacob disease. Kuru is a fatal infection resulting from degeneration of the brain. It is limited to New Guinea cannibals. Creutzfeldt-Jakob disease is a degenerative disorder of the nervous system which always ends in death. Initial symptoms include forgetfulness, loss of coordination, spasticity, coma and paralysis. This disease is classified as a slow virus disease although no virus has as yet been identified. The major source of infection is from ingestion of brains from animals infected with the disease.

(h) Mononucleosis is a viral infection but the virus has not yet been isolated and no treatment is available. Symptoms in adults include sore throat, fatigue, fever, swelling of lymph glands and an increase in mononuclear white blood cells. The infection is usually mild but if the spleen becomes enlarged there is a danger of rupture and subsequent death. The route of transmission is primarily by oral contact. Most symptoms manifest themselves about a week after infection; duration is about two to three weeks.

(i) Rabies is a fatal disease unless treated early. It is acquired through the bite of an infected animal, symptoms include headache, loss of appetite, fever, nausea, sensitivity to loud noises and bright lights, muscle spasms in the throat, seizures, paralysis, and finally death. These symptoms result from destruction of brain tissue which is clearly evident under the microscope. Treatment involves daily injection with a vaccine.

(j) AIDS is the acronym for acquired immune deficiency syndrome, a disorder characterized by a breakdown of the body's immunity resulting in susceptibility to certain infections and rare forms of cancer. The AIDS virus also produces milder effects known as AIDS-related complex (ARC). Symptoms include chronic fatigue, weight loss, and enlargement of lymph nodes. Most people infected by the virus, however, are asymptomatic. About 25 percent of those with positive blood tests for the virus develop AIDs, another 25 percent develop ARC.

The virus causing AIDS is a retrovirus known as human immunodeficiency virus (HIV). It is present not only in blood but also in vaginal secretions and sperm. The virus is transmitted primarily by sexual contact and through use of contaminated needles. The virus can be transmitted prior to birth or during vaginal delivery and may also be transmitted in breast milk.

Newborns who acquire AIDs in utero are born with low birthweights, enlarged livers and spleens and after birth do not grown as fast as other children. Many develop the same clinical problems as adults such as increased susceptibility to infection. Neurological complications include delayed motor development, below normal intelligence and atrophy of areas of the brain in some cases.

<u>Rubella</u> is an RNA reovirus that causes an illness called rubella or German measles. In the adult symptoms are mild and mostly inconsequential such as a skin rash, sore throat, runny nose, conjunctivitis, and coughing. However, in some cases infection may result in multiple sclerosis and systemic lupus erythematosus. In young children pneumonia and inner ear infection are also common. The most serious consequence for children and adults is encephalitis.

Infection is by way of nose or mouth, often by coughing. The virus multiplies in the lymph glands, enters the blood and is distributed throughout the body, eventually concentrating on the respiratory organs, eyelids and skin.

The fetus rubella virus can produce a variety of effects, some of which are nonspecific such as low birthweight and enlarged spleen. Symptoms typically occur in combination except for deafness which often appears as the sole defect. A classic triad of effects, consisting of sensorineural hearing loss, cataracts, and congenital heart disease is also very common. The hearing loss is due to damage to the organ of Corti. Loss of hearing may be a reason for delayed language development. The most common heart disorder is patent ductus arteriosus but peripheral pulmonary artery stenosis, and atrial and ventricular septal defects are also common. Other effects include spontaneous abortion, stillbirth, low birth weight, retinitis, enlarged liver and spleen and lesions in bones.

The time of exposure is critical for type of defect. Infection during the first 8 weeks of gestation results in the highest incidence of defects and the greatest likelihood of multiple defects. Infection at any time during the first 16 weeks can result in hearing loss but if infection occurs just before the 16th week, deafness is usually the only result. After 16 weeks, intrauterine transmission decreases markedly.

Although the panorama of effects has been well characterized, the mechanism by which the virus produces these effects is not well known. Some evidence suggests interference with normal cell division resulting in a disordering of the timetable for developmental processes.

Around 3 to 6 months of age other effects begin to appear including failure to thrive, pneumonia, skin rashes, meningitis, and neurological damage. About half of those infected in utero are mentally retarded.

Multiple sclerosis is believed by some to result from an autoimmune response to the rubella virus which attacks the brain. The disorder is often found in people with high levels of antibodies to the measles virus and may in fact be a slow virus type of measle infection. Symptoms include impairment of vision, speech, arm and leg movements, dizziness, loss of bowel and bladder control, muscle spasm, paralysis and death. No cure is available as yet but drugs can be administered to reduce the severity of symptoms and slow down its course of action.

In newborns with a rash, detection is relatively straightforward. Laboratory confirmation relies on isolation of the virus from the throat and its subsequent culturing or by an increase in blood borne antibodies.

In 1969, a live attenuated rubella virus vaccine became available. Prior to its introduction about 75% of adults became infected by late adulthood, very often during periodic epidemics which occurred every 6 to 9 years. The vaccine greatly reduced infection rates especially in young children and most cases now occur among adolescents and adults of childbearing age. Reinfection may occur among those with natural and vaccine-induced immunity but these reinfections do not affect the fetus.

To reduce the risk of intrauterine transmission, it is now standard obstetrical practice to determine the immune status of the patient and routine blood tests are available for diagnosis.

Cytomegalovirus is the most common cause of intrauterine infections. The morbidity rate is also the highest and far exceeds that of rubella, although its effects have not been appreciated to the same extent. The reason for this apparent neglect is that the symptoms are often "silent" at birth, i.e., manifestations of infection usually are not expressed until some time after birth. Cytomegalovirus is a member of the herpes virus family. The name arose from the appearance of large cells ("cyto-," cell and "mega-," large) in the internal organs of stillborn infants. Initially, these cells were believed to be parasites but eventually they were identified as "inclusions"--dark organelles inside the cell that accompany infection. About 50% of all adults in the U.S. have antibodies to cytomegalovirus. The infection is transmitted by sexual contact and by blood. Primary infection in adults is asymptomatic although a mononucleosis syndrome may occur. After primary infection, the virus persists in a latent phase and may be reactivated during times when the immune system is surpressed such as occurs during pregnancy.

Intrauterine transmission of cytomegalovirus is a major cause of premature death and birth defects. Infection occurs in about 0.5 to 2.5% of all live births. Usually the infected infant is asymptomatic at birth. Asymptomatic infants also tend to perform well on developmental tests in early life. When symptoms do become observable, they are characterized by enlargement of liver, spleen and jaundice. The kidney and lung are also damaged. In many cases the health of the infected neonate rapidly deteriorates and death occurs within weeks of birth. Most infants who succumb are premature or small for date. Survivors experience central nervous system (CNS) defects including microcephaly, psychomotor retardation, blindness, and mental retardation. Cytomegalovirus is also one of the major nonhereditary causes of deafness. About 80% of infected children develop CNS problems regardless of symptoms at birth. Since symptomatology in the adult is mild, prenatal infection may not be noticed. Currently no vaccines are available for prevention. Detection of infection relies on observation of inclusion-containing cells.

The route of transmission is still unknown. In some cases it appears that the critical factor is presence of the virus in maternal blood during the first trimester of pregnancy since many of the abnormalities are compatible with infection at this period of development. However, it is often not possible to detect the virus in maternal blood although its presence in the cervix is detectable.

Cytomegalovirus infection is widespread. About 80% of all the population is infected at some time of their lives. Direct physical contact is required either by mouth to mouth contact or by blood in organ transfusion. Infection is especially high where personal hygiene is poor.

Herpes Simplex Virus (HSV) is a general term for two related sexually transmitted viruses. Herpes simplex type I (HSV-I) infection is transmitted by orolabial contact. In adults infection is associated most often with recurrent cold sores or fever blisters on the upper lip. Other symptoms in the adult are fever, swelling of lymph glands, inflammation of the mouth, and malaise. In some cases, more severe symptoms such as encephalitis and meningitis occur resulting in facial paralysis (Bell's palsy), convulsions and death. The eye is another site of symptomatic infection but this occurs much more infrequently than the lip. However, the consequences are much more severe including cold sores on the eyelid, inflammation of the conjunctival membrane, and possible blindness.

Herpes simplex type II (HSV-II) is similar to herpes HSV-I except the signs of infection occur on or around the genitals (hence its other designation as genital herpes). The virus is spread through sexual intercourse and about 10 to

30% of women of childbearing age have specific antibody to HSV-II. Herpes type II is now considered the second most common veneral disease, the first being gonorrhea. Symptoms include inflammation of the genital area, swelling of lymph glands in the groin, and fever.

Fetal infection by herpes occurs primarily during vaginal delivery. If a first attack occurs late in pregnancy there may not be time enough for antibodies to form and cross the placenta to protect the infant at term. Since there is a possibility of infection from vaginal delivery, one way of circumventing this route of infection is by cesearian delivery. However, many women who are asymptomatic carry the virus resulting in a high rate of infection.

Neonates infected with herpes are often born premature and experience a number of complications, although in most cases these do not become apparent until several weeks after birth. Typically, the first sign of infection is skin lesions. In some cases the infection remains localized in the skin or eye but in the majority of cases the infection spreads to internal organs such as liver, spleen and brain. Indications of dissemination are lethargy, poor feeding, pneumonia, hepatitis, hypotension and sometimes seizures. Infection of the brain is especially associated with focal motor seizures, poor response to anticonvulsive therapy, high pitched crying, poor reflex tone, hypothermia and apnea.

# VIRUSES AND REPRODUCTION

# A

1. Aase, J.M., Noren, G.R., Reddy, D.V. and Geme, J.W., Jr. Mumps-virus infection in pregnant women and the immunologic response of their offspring. N. Engl. J. Med., 1972, 286:1379-1382.

2. Aaskov, J. Absence of intrauterine infection following Ross River virus infection during pregnancy. Am. J. Trop. Med. Hyg., 1984, 33:739-740.

3. Aaskov, J.G., Davies, C.E., Tucker, M. and Dalglish, D. Effect on mice of infection during pregnancy with three Australian arboviruses. Am. J. Trop. Med. Hyg., 1981, 30:198-203.

4. Aaskov, J.G., Nair, K., Lawrence, G.W., Dalglish, D.A. and Tucker, M. Evidence for transplacental transmission of Ross River virus in humans. Med. J. Aust., 1981, 2:20-21.

5. Abdel, Aziz, A.A., Meegan, J.M. and Laughlin, L.W. Rift Valley fever as a possible cause of human abortions. Trans. R. Soc. Trop. Med. Hyg., 1980, 74:685-686.

6. Abe, S. Effect of immunization with fetal cells on adenovirus-12 oncogenesis. Acta. Med. Okayama, 1974, 28:147-158.

7. Ablin, R.J. and Curtis, W.W. Treatment of condylomata acuminata. S. Afr. Med. J., 1973, 47:1442.

8. Abramson, H., Greenberg, M. and Magee, M.C. Poliomyelitis in the newborn infant. J. Pediat., 1953, 43:167.

9. Ackermann, O. Early immunization against canine distemper and hepatitis, using combined vaccines. J. Am. Vet. Med. Assoc., 1970, 156:1755-1758.

10. Ackermann, R., Kalter, S.S., Heberling, R.L., McCullough, B., Eichberg, J. and Rodriguez, A.R. Fetal infection of the baboon (Papio cynocephalus) with lymphocytic choriomeningitis virus. Arch. Virol., 1979, 60:311-323.

11. Adams, W.H., Shrestha, S.M. and Adams, D.G. Coagulation studies of viral hepatitis occurring during pregnancy. Am. J. Med. Sci., 1976, 272:139-145.

12. Adelstein, A.M. and Donovan, J.W. Malignant disease in children whose mothers had chickenpox, mumps, or rubella in pregnancy. Br. Med. J., 1972, 4:629-631.

13. Adkisson, M.A. Herpes zoster in a newborn premature infant. J. Ped., 1965, 66:956.

14. Adler, S.P., Chandrika, T., Lawrence, L. and Baggett, J. Cytomegalovirus infections in neonates acquired by blood transfusions. Pediatr. Infect. Dis., 1983, 2:114-118.

15. Afshar, A. A review of non-bite transmission of rabies virus infection. Br. Vet. J., 1979, 135:142-148.

16. Agarwal, N.R., Bardoloi, J.N., Sehgal, S. and Gupta, A.N. Placental transmission of neutralizing antibodies to poliomyelitis virus. Indian J. Med. Res., 1971, 59:1703-1707.

17. Agrawal, N.R., Singh, V., Bhattacharya, S.K., Sundar, S. and Tiwari, P. Prognostic significance of zinc in pregnancy with hepatitis. Asia Oceania J. Obstet. Gynaecol., 1985, 11:589-591.

18. Aguilar, M.J., Calanchini, P.R. and Finley, K.H. Perinatal arbovirus encephalitis and its sequelae. Res. Publ. Assoc. Res. Nerv. Ment. Dis., 1968, 44:216-235.

19. Ahlfors, K. Epidemiological studies of congenital cytomegalovirus infection. Scand. J. Infect. Dis., 1982, 34:1-36.

20. Ahlfors, K., Forsgren, M., Ivarsson, S.A., Harris, S. and Svanberg, L. Congenital cytomegalovirus infection: on the relation between type and time of maternal infection and infant's symptoms. Scand. J. Infect. Dis., 1983, 15:129-138.

21. Ahlfors, K., Harris, S., Ivarsson, S. and Svanberg, L. Secondary maternal cytomegalovirus infection causing symptomatic congenital infection, N. Eng. J. Med., 1981, 305:284.

22. Ahlfors, K., Ivarsson, S.A. and Bjerre, I. Microcephaly and congenital cytomegalovirus infection a combined prospective and retrospective study of a Swedish infant population. Pediatrics, 1986, 78:1058-1063.

23. Ahlfors, K., Ivarsson, S.A., Harris, S., Svanberg, L., Holmqvist, R., Lernmark, B. and Theander, G. Congenital cytomegalovirus infection and disease in Sweden and the relative importance of primary and secondary maternal infections. Preliminary findings from a prospective study. Scand. J. Infect. Dis., 1984, 16:129-137.

24. Ahlfors, K., Ivarsson, S.A., Johnsson, T. and Svanberg, L. A prospective study on congenital and acquired cytomegalovirus infections in infants. Scand. J. Infect. Dis., 1979, 11:177-178.

25. Ahlfors, K., Ivarsson, S.A., Johnsson, T. and Svanberg, L. Primary and secondary maternal cytomegalovirus infections and their relation to congenital infection. Analysis of maternal sera. Acta. Paediatr. Scand., 1982, 71:109-113.

26. Ahmed, M., Schidlovsky, G., Korol, W., Vidrine, G. and Cicmanec, J.L. Occurrence of Mason-Pfizer monkey virus in healthy rhesus monkeys. Cancer Res., 1974, 34:3504-3508.

27. Aho, K., Cantell, K., Hjelt, L. and Rantasalo, I. Possible association between maternal virus infections and stillbirths. Br. J. Prev. Soc. Med., 1968, 22:38-40.

28. Aickin, D.R. and Campbell, D.G. Anomalous amniotic fluid bilirubin after hepatitis in an Rh immunized patient. Obstet. Gynecol., 1971, 37:687-688.

29. Aitkens, G.H., Bowman, R. and Hansman, D. A case of foetal vaccinia. Med. J. Aust., 1968, 2:173-174.

30. Akagi, T. Fetal antigens associated with adenovirus 12 mouse tumors. Acta. Med. Okayama., 1976, 30:385-395.

31. Akhtar, K.A. and Akhtar, M.A. Viral hepatitis in pregnancy. J.P.M.A., 1979, 29:31-35.

32. Alagille, D. Vertical transmission of HBV hepatitis to fetuses and infants: current advances in prevention. J. Pediatr. Gastroenterol. Nutr., 1985, 4:515-517.

33. Albrecht, P., Ennis, F.A., Saltzman, E.J. and Krugman, S. Persistence of maternal antibody in infants beyond 12 months: mechanism of measles vaccine failure. J. Pediatr., 1977, 91:715-718.

34. Albrecht, T. and Rapp, F. Malignant transformation of hamster embryo fibroblasts following exposure to ultraviolet-irradiated human cytomegalovirus. Virology, 1973, 55:53-61.

35. Aleck, K.A., Rosen, L., Pettitt, D.J., Boveington, C. and Bennett, P.H. Absence of intrauterine infection following Ross River virus infection during pregnancy. Am. J. Trop. Med. Hyg., 1983, 32:618-620.

36. Alessandro, Z.R., Elisabetta, T., Pierino, F. and Enrico, M. Vertical transmission of the HBV-associated delta agent. Prog. Clin. Biol. Res., 1983, 143:127-132.

37. Alexander, E.R. Maternal and neonatal infection with cytomegalovirus in Taiwan. Pediatr. Res., 1967, 1:210-211.

38. Alexander, G.J. and Eddleston, A.L. Does maternal antibody to core antigen prevent recognition of transplacental transmission of hepatitis-B-virus infection? Lancet, 1986, 8476:296-297.

39. Alexander, I. Congenital varicella. Br. Med. J., 1979, 2:1074.

40. Alexandersen, S. Acute interstitial pneumonia in mink kits: experimental reproduction of the disease. Vet. Pathol., 1986, 23:579-588.

41. Alfano, J.E. Ocular aspects of the maternal rubella syndrome. Trans. Am. Acad. Ophthalmol. Otolaryngol., 1966, 70:235-266.

42. Alfano, J.E. Maternal virus infections and fetal ocular malformation. Postgrad. Med., 1969, 45:113-116.

43. Alford, B.R. Rubella -- a challenge for modern medical science. Arch. Otolaryngol., 1968, 88:1-2.

44. Alford, B.R. Rubella - la bete noire de la medecine. Laryngoscope, 1968, 78:1623-1659.

45. Alford, C.A., Jr. Studies on antibody in congenital rubella infections. I. Physicochemical and immunologic investigations of rubella neutralizing antibody. Am. J. Dis. Child., 1965, 110:455-643.

46. Alford, C.A., Jr. Congenital rubella: a model for chronic in utero infections in man. Ala. J. Med. Sci., 1966, 3:463-472.

47. Alford, C.A., Jr. Congenital rubella. A review of the virologic and serologic phenomena occurring after maternal rubella in the first trimester. Southern Med. J., 1966, 59:745.

48. Alford, C.A., Jr. and Kanich, L.S. Congenital rubella: a review of the virologic and serologic phenomena occurring after maternal rubella in the first trimester. South Med. J., 1966, 59:745-748.

49. Alford, C.A., Schaefer, J., Blankenship, W.J., Straumfjord, J.V. and Cassady, G. A correlative immunologic, microbiologic and clinical approach to the diagnosis of acute and chronic infections in newborn infants. N. Engl. J. Med., 1967, 277:437-449.

50. Alford, C.A., Stagno, S. and Pass, R.F. Perinatal Infections: Natural history of perinatal cytomegaloviral infection. Ciba. Found. Symp., 1980, 125-147.

51. Alford, C.A., Stagno, S. and Reynolds, D.W. Long term mental and perceptual defects associated with silent intrauterine infections, in Gluck, L. (ed): Intrauterine Asphyxia and the Developing Fetal Brain, Chicago, Year Book Medical Publishers, 1977, 377-393.

52. Allan, B.C. Immunization against rubella. Med. J. Aust., 1971, 1:1406-1407.

53. Allan, B.C., Hamilton, S.M., Wiemers, M.A., Winsor, H. and Gust, I.D. Pregnancy complicated by accidental rubella vaccination. Aust. NZ. J. Obstet. Gynaecol., 1973, 13:72-76.

54. Allan, P.J., Dennett, D.P. and Johnson, R.H. Studies on the effects of infectious bovine rhinotracheitis virus on reproduction in heifers. Aust. Vet. J., 1975, 51:370-373.

55. Allen, G.P., Yeargan, M.R., Turtinen, L.W. and Bryans, J.T. A new field strain of equine abortion virus (equine herpesvirus-1) among Kentucky horses. Am. J. Vet. Res., 1985, 46:138-140.

56. Allen, J.C., Watters, G., OGorman, A.M., Patriquin, H., Meloff, K., Middleton, P. and Armstrong, D. Herpes virus type 2 encephalitis in two neonates from the same household. J. Pediatr., 1976, 89:949-950.

57. Allison, A.C. and Patton, G.R. Chromosomal abnormalities in human diploid cells infected with mycoplasma and their possible relevance to aetiology of Down's syndrome (mongolism). Lancet, 1966, 2:1229.

58. Altemeier, W.A., III, Mundon, F.K., Top, F.H., Jr. and Russell, P.K. Method for extracting viral hemagglutination-inhibiting antibodies from the nonspecific inhibitors of serum. Appl. Microbiol., 1970, 19:785-790.

59. Alter, M. and Schulenberg, R. Dermatoglyphics in the rubella syndrome. J.A.M.A., 1966, 197:685-688.

60. Altshuler, G. Implications of two cases of human placental plasma cells. Reports of cytomegalic inclusion disease in a thirteen week's fetus and of ascending herpes infection in the newborn. Am. J. Pathol., 1973, 70:18a.

61. Altshuler, G. Immunologic competence of the immature human fetus. Morphologic evidence from intrauterine Cytomegalovirus infection. Obstet. Gynecol., 1974, 43:811-816.

62. Altshuler, G. Placentitis, with a new light on an old TORCH. Obstet. Gynecol. Annu., 1977, 6:197-221.

63. Altshuler, G. Case 2. Placenta of erythroblastosis and maternal hepatitis. Pediatr. Pathol., 1984, 2:337-338.

64. Altshuler, G. and McAdams, A.J. Cytomegalic inclusion disease of a nineteen-week fetus. Case report including a study of the placenta. Am. J. Obstet. Gynecol., 1971, 111:295-298.

65. Alysworth, A.A. and Monif, G.R. Delayed serologic evidence of infection with rubella virus after the administration of gammaglobulin. A case report. Obstet. Gynecol., 1971, 38:752-754.

66. Ambrose, K.R., Anderson, N.G. and Coggin, J.H. Interruption of SV40 oncogenesis with human foetal antigen. Nature, 1971, 233:194-195.

67. Ambrus, M., Peley, I., Bajtai, G. and Pacsa, S. Congenital rubella syndrome: clinical, virological and immunological aspects. Haematologia, 51:117-126.

68. Ames, M.D., Plotkin, S.A., Winchester, R.A. and Atkins, T.E. Central auditory imperception: a significant factor in congenital rubella deafness. J.A.M.A., 1970, 213:419-421.

69. Ammann, A.J., Cowan, M.J., Wara, O.W., Weintrub, P., Dritz, S., Goldman, H. and Perkins, H.A. Acquired immunodeficiency in an infant: possible transmission by means of blood products. Lancet, 1983, 1:956-958.

70. Amstey, M.S. Intra-amniotic inoculation of rubella virus. Am. J. Obstet. Gynecol., 1969, 104:573-577.

71. Amstey, M.S. Management of pregnancy complicated by genital herpes virus infection. Obstet. Gynecol., 1971, 37:515-520.

72. Amstey, M.S. Letter: Fetal hepatitis. Obstet. Gynecol., 1974, 43:794-795.

73. Amstey, M.S. Letter: Rubella virus transport in pregnancy. J.A.M.A., 1974, 228:696.

74. Amstey, M.S. Viral infection in pregnancy: some general principles. Clin. Obstet. Gynecol., 1974, 19:55-60.

75. Amstey, M.S. Neonatal risk following late gestational genital herpesvirus hominis infection. Am. J. Dis. Child, 1975, 129:985.

76. Amstey, M.S. Maternal viral infection with adverse results: cytomegalovirus and herpesvirus. Semin. Perinatol., 1977, 1:1-10.

77. Amstey, M.S. Vaccination in pregnancy. Clin. Obstet. Gynaecol., 1983, 10:13-22.

78. Amstey, M.S. Transplacental spread of influenza virus. Am. J. Obstet. Gynecol., 1985, 151:828-829.

79. Amstey, M.S. and Monif, G.R. Genital herpesvirus infection in pregnancy. Obstet. Gynecol., 1974, 44:394-397.

80. Amstey, M.S., Monif, G.R., Nahmias, A.J. and Josey, W.E. Cesarean section and genital herpesvirus infection. Obstet. Gynecol., 1979, 53:641-642.

81. Amstey, M.S., Nasello, M.A. and Hochberg, C.J. Multiple virus isolations from a pregnant patient. Am. J. Obstet. Gynecol., 1973, 115:897-900.

82. Anand, A., Gray, E.S., Brown, T., Clewley, J.P. and Cohen, B.J. Human parvovirus infection in pregnancy and hydrops fetalis. N. Engl. J. Med., 1987, 316:183-186.

83. Anders, B.J., Lauer, B.A. and Foley, L.C. Computerized tomography to define CNS involvement in congenital cytomegalovirus infection. Am. J. Dis. Child, 1980, 134:795-797.

84. Andersen, A.A. and Campbell, C.H. Experimental placental transfer of foot-and-mouth disease virus in mice. Am. J. Vet. Res., 1976, 37:585-589.

85. Andersen, A.A. and Campbell, C.H. Experimental placental transfer of Akabane virus in the hamster. Am. J. Vet. Res., 1978, 39:301-304.

86. Andersen, A.A. and Hanson, R.P. Intrauterine infection of mice with St. Louis encephalitis virus: immunological, physiological, neurological, and behavioral effects on progeny. Infect. Immun., 1975, 12:1173-1183.

87. Andersen, H.K., Brostrom, K., Hansen, K.B., Leerhoy, J., Pedersen, M., Osterballe, O., Felsager, U. and Mogensen, S. A prospective study on the incidence and significance of congenital cytomegalovirus infection. Acta. Paediatr. Scand., 1979, 68:329-336.

88. Andersen, H.K., Gravesen, J.J. and Iversen, T. Cytomegalovirus infection among infants admitted to a paediatric department. Acta. Paediatr. Scand., 1972, 61:445-451.

89. Andersen, H.K. and Jeppesen, T. Virus - like particles in guinea pig oogonia and oocytes. J. Natl. Cancer Inst., 1972, 49:1403-1410.

90. Anderson, B.C., Bulgin, M.S., Adams, S. and Duelke, B. Firm udder in periparturient ewes with lymphocytic accumulations, retrovirus infection, and milk unavailable at the teat. J. Am. Vet. Med. Assoc., 1985, 186:391-393.

91. Anderson, C.K. and Jensen, R. Pathologic changes in placentas of ewes inoculated with bluetongue virus. Am. J. Vet. Res., 1969, 30:987-989.

92. Anderson, G.W., Anderson, G., Skaar, A. and Sandler, F. Poliomyelitis in pregnancy. Amer. J. Hyg., 1952, 55:127.

93. Anderson, H., Barr, B. and Wedenberg, E. Genetic disposition -- a prerequisite for maternal rubella deafness. Arch. Otolaryngol., 1970, 91:141-147.

94. Anderson, J.M. and Nicholls, M.W. Herpes encephalitis in pregnancy. Br. Med. J., 1972, 1:632.

95. Anderson, K.E., Stevens, C.E., Tsuei, J.J., Lee, W.C., Sun, S.C. and Beasley, P. Hepatitis B antigen in infants born to mothers with chronic hepatitis B antigenemia in Taiwan. Am. J. Dis. Child, 1975, 129:1389-1392.

96. Anderson, O.W. Mumps mastitis. J. Pediatr., 1977, 91:687.

97. Andreasson, B., Svenningsen, N.W. and Nordenfelt, E. Screening for viral infections in infants with poor intrauterine growth. Acta. Paediatr. Scand., 1981, 70:673-676.

98. Andres, J.M. and Walker, W.A. Effect of Escherichia coli endotoxin on the developing rat liver. I. Giant cell inducation and disruption in protein metabolism. Pediatr. Res., 1979, 13:1290-1293.

99. Andrews, C.A., Daniel, R.W. and Shah, K.V. Serologic studies of papovavirus infections in pregnant women and renal transplant recipients. Prog. Clin. Biol. Res., 1983, 105:133-141.

100. Anglin, C. Varicella and congenital defects. Can. Med. Assoc. J., 1973, 108:959.

101. Ansari, B.M. and Mason, M.K. Chromosomal abnormality in congenital rubella. Pediatrics, 1977, 59:13-15.

102. Antoine, C., Morris, M. and Douglas, G. Maternal and fetal mortality in acquired immunodeficiency syndrome. N.Y. State J. Med., 1986, 86:443-445.

103. Appel, M., Carmichael, L.E. and Robson, D.S. Canine adenovirus type 2-induced immunity to two canine adenoviruses in pups with maternal antibody. Am. J. Vet. Res., 1975, 36:1199-1202.

104. Armitage, J.O., Feagler, J.R. and Skoog, D.P., Burkitt lymphoma during pregnancy with bilateral breast involvement. J.A.M.A., 1977, 237:151.

105. Armon, P.J. Burkitt's lymphoma of the ovary in association with pregnancy. Two case reports. Br. J. Obstet. Gynaecol., 1976, 83:169-172.

106. Arora, B. Infective hepatitis and pregnancy in Bristol. J. Obstet. Gynaecol. Br. Commonw., 1967, 74:763-765.

107. Arvin, A.M., Hensleigh, P.A., Prober, C.G., Au, D.S., Yasukawa, L.L., Wittek, A.E., Palumbo, P.E., Paryani, S.G. and Yeager, A.S. Failure of antepartum maternal cultures to predict the infant's risk of exposure to herpes simplex virus at delivery. N. Engl. J. Med., 1986, 315:796-800.

108. Asse, J.M., Noren, G.R., Reddy, D.V. and St. Geme, J.W. Mumps-virus infection in pregnant women and the immunological response of their offspring. N. Engl. J. Med., 1972, 286:1379.

109. Atkins, G.J., Carter, J. and Sheahan, B.J. Effect of alphavirus infection on mouse embryos. Infect. Immun., 1982, 38:1285-1290.

110. Auruskin, T.W., Brakin, M. and Juan, C. Congenital myxoedema and congenital rubella. Pediatrics, 1982, 69:495-496.

111. Austin, D.F., Karp, S., Dworsky, R. and Henderson, B.E. Excess leukemia in cohorts of children born following influenza epidemics. Am. J. Epidemiol., 1975, 101:77-83.

112. Avery, G.B., Monif, G.G., Sever, J.L. and Leikin, S.L. Rubella syndrome after inapparent maternal illness. Am. J. Dis. Child, 1965, 110:444-446.

113. Aycock, W.L. The frequency of poliomyelitis in pregnancy. N. Eng. J. Med., 1941, 225:405.

114. Aycock, W.L. Acute poliomyelitis in pregnancy: Its occurrence according to month of pregnancy and sex of fetus. N. Eng. J. Med., 1946, 235:160.

115. Aycock, W.L. and Ingalls, T.H. Maternal disease as a principle in the epidemiology of congenital anomalies. Amer. J. Med. Sci., 1946, 212:366.

116. Ayoola, E.A., Ogunbode, O. and Odelola, H.A. Congenital transmission of hepatitis B antigen in Nigerians. Arch. Virol., 1981, 67:97-99.

# B

117. Babb, R.R. Hepatitis B antigen: a review of its importance in the practice of obstetrics and gynecology. Am. J. Obstet. Gynecol., 1976, 126:602-606.

118. Bader, M. Polio vaccination during pregnancy. J.A.M.A., 1983, 249:2018-2019.

119. Badman, R.T., Mitchell, G., Jones, R.T. and Westbury, H.A. Association of bovine viral diarrhoea virus infection to hydranencephaly and other central nervous system lesions in perinatal calves. Aust. Vet. J., 1981, 57:306-307.

120. Bahrani, M., Boxerbaum, B., Gilger, A.P., Rosenthal, M.S. and Teree, T.M. Generalized herpes simplex and hypoadrenocorticism. A case associated with adrenocortical insufficiency in a prematurely born male: clinical, virologic, ophthalmological, and metabolic studies. Am. J. Dis. Child, 1966, 111:437-445.

121. Bai, P.V. and John, T.J. Congenital skin ulcers following varicella in late pregnancy. J. Pediatr., 1979, 94:65-67.

122. Baker, D.A., Bienstock, J., Metz, G., Rochelson, B.L. and Monheit, A.G. Serologic screening of pregnant women at high risk for transmitting hepatitis B to their newborn. Bull. N.Y. Acad. Med., 1986, 62:282-286.

123. Baker, D.A. and Phillips, C.A. Maternal and neonatal infection with coxsackievirus. Obstet. Gynecol., 1980, 55:12-15.

124. Baker, D.A. and Plotkin, S.A. Enhancement of vaginal infection in mice by herpes simplex virus type II with progesterone. Proc. Soc. Exp. Biol. Med., 1978, 158:131-134.

125. Baker, D.A. and Plotkin, S.A. Herpes simplex virus type 2 infection of neonatal mice. Infection, 1978, 6:158-161.

126. Baker, D.A. and Plotkin, S.A. Genital herpes simplex virus (HSV) isolation during pregnancy. Obstet. Gynecol., 1979, 53:9-12.

127. Baker, D.A. and Polk, B.F. Hepatitis B: a controllable disease. Obstet. Gynecol., 1983, 62:105-108.

128. Balfour, H.H., Jr. Prevention of viral infections during pregnancy. Clin. Obstet. Gynecol., 1982, 25:619-626.

129. Balistreri, W.F. Viral hepatitis: implications to pediatric practice. Adv. Pediatr., 1985, 32:287-320.

130. Ball, R.F., Hill, J.F., Lyman, J. and Wyatt, A. The effect of early natural exposure to Marek's disease on immunization of breeders by vaccination. Poult. Sci., 1971, 50:648-649.

131. Ballard, R.A., Drew, W.L., Hufnagle, K.G. and Reidel, P.A. Acquired cytomegalovirus infection in preterm infants. Am. J. Dis. Child, 1979, 133:482-485.

132. Banatvala, J.E. Laboratory investigations in the assessment of rubella during pregnancy. Br. Med. J., 1968, 1:561-562.

133. Banatvala, J.E. Maternal rubella and its virological diagnosis. Postgrad. Med. J., 1972, 48:11-17.

134. Banatvala, J.E. Rubella -- continuing problems. Br. J. Obstet. Gynaecol., 1985, 92:193-196.

135. Banatvala, J.E., Best, J.M., Bertrand, J., Bowern, N.A. and Hudson, S.M. Serological assessment of rubella during pregnancy. Br. Med. J., 1970, 1:247-250.

136. Banatvala, J.E., Horstmann, D.M., Payne, M.C. and Gluck, L. Rubella syndrome and thrombocytopenic purpura in newborn infants. Clinical and virologic observations. N. Engl. J. Med., 1965, 273:474-478.

137. Banatvala, J.E., Potter, J.E. and Best, J.M. Interferon response to sendai and rubella viruses in human foetal cultures, leucocytes and placental cultures. J. Gen. Virol., 1971, 13:193-201.

138. Banatvala, J.E., Potter, J.E. and Webster, M.J. Foetal interferon responses induced by rubella virus. Ciba. Found. Symp., 1972, 10:77-99.

139. Banerji, D. Possible congenital infection with kala-azar. J. Indian Med. Assoc., 1955, 24:433.

140. Banks, R.A., Babcock, G.F., Whitmore, A.C. and Haughton, G. Effect of immunologic intervention on in vivo murine Rous sarcoma virus tumorigenesis. J.N.C.I., 1979, 63:1423-1431.

141. Barbosa, L.H., London, W.T., Hamilton, R. and Buckler, C. Interferon response of the fetal Rhesus monkey after viral infection. Proc. Soc. Exp. Biol. Med., 1974, 146:398-400.

142. Barin, F., Goudeau, A., Denis, F., Yvonnet, B., Chiron, J.P., Coursaget, P. and Mar, I.D. Immune response in neonates to hepatitis B vaccine. Lancet, 1982, 1:251-253.

143. Barlow, R.M. Morphogenesis of hydranencephaly and other intracranial malformations in progeny of pregnant ewes infected with pestiviruses. J. Comp. Pathol., 1980, 90:87-98.

144. Barlow, R.M. Some interactions of virus and maternal/foetal immune mechanisms in border disease of sheep. Prog. Brain Res., 1983, 59:255-268.

145. Barlow, R.M., Rennie, J.C., Gardiner, A.C. and Vantsis, J.T. Infection of pregnant sheep with the NADL strain of bovine virus diarrhoea virus and their subsequent challenge with Border disease IIB pool. J. Comp. Pathol., 1980, 90:67-72.

146. Barnes, R.D. Studies on transfer of MuLV and lymphoma development. In: Chandra P., ed. Antiviral mechanisms in the control of neoplasia. New York, Plenum Press, 1979, 67-80.

147. Barnett, J.B., Holcomb, D., Menna, J.H. and Soderberg, L.S. The effect of prenatal chlordane exposure on specific anti-influenza cell-mediated immunity. Toxicol., 1985, 25:229-238.

148. Barno, A. Terminating pregnancy after first-trimester rubella. Postgrad. Med., 1969, 46:177-179.

149. Baron, J., Youngblood, L., Siewers, C.M.F. and Medearis, D.N. Jr. The incidence of cytomegalovirus, herpes simplex, rubella, and toxoplasma antibodies in microcephalic mentally retarded and normocephalic children. Pediatrics, 1969, 44:932-939.

150. Barr, B. and Lundstrom, R. Deafness following maternal rubella. Retrospective and prospective studies. Acta. Otolaryngol. (Stockh.), 1961, 53:413.

151. Barr, S.J. More hydatidiform moles? Br. Med. J., 1977, 2:707-708.

152. Barsegar, B., Hofmann, H. and Zweymuller, E. The diagnosis of measles in the newborn and infant age groups. Z. Kinderheilkd., 1972, 113:175-184.

153. Bart, S.W., Stetler, H.C., Preblud, S.R., Williams, N.M., Orenstein, W.A., Bart, K.J., Hinman, A.R. and Herrmann, K.L. Fetal risk associated with rubella vaccine: an update. Rev. Infect. Dis., 1985, 1:95-102.

154. Bartha, A. and Mate, S. Transplacental transmission of bovine adenoviruses. Comp. Immunol. Microbiol. Infect. Dis., 1983, 6:189-192.

155. Basalamah, A.H., Serebour, F. and Kazim, E. Materno-foetal transmission of hepatitis B in Saudi Arabia. J. Infect., 1984, 8:200-204.

156. Basar, T., Bhargava, V.L., Pande, Y. and Seth, P. Prevalence of genital herpes virus (HSV) infection in pregnancy. Indian J. Med. Res., 1984, 79:327-332.

157. Baskar, J.F. and Huang, E.S. Effect of murine cytomegalovirus on implantation stage mouse embryos. Proc. Soc. Exp. Biol. Med., 1981, 167:581-587.

158. Baskar, J.F., Stanat, S.C. and Huang, E.S. Congenital defects due to reactivation of latent murine cytomegaloviral infection during pregnancy. J. Infect. Dis., 1985, 152:621-624.

159. Baskar, J.F., Stanat, S.C., Sulik, K.K. and Huang, E.S. Murine cytomegalovirus-induced congenital defects and fetal maldevelopment. J. Infect. Dis., 1983, 148:836-843.

160. Bassing, S. Saving the baby when Mom has herpes. R.N., 1985, 48:35-37.

161. Batcup, G., Holt, P., Hambling, M.H., Gerlis, L.M. and Glass, M.R. Placental and fetal pathology in Coxsackie virus A9 infection: a case report. Histopathology, 1985, 9:1227-1235.

162. Bates, H.R., Jr. Coxsackie virus B3 calcific pancarditis and hydrops fetalis. Am. J. Obstet. Gynecol., 1970, 106:629-630.

163. Bates, T. Poliomyelitis in pregnancy, fetus, and newborn. Amer. J. Dis. Child, 1955, 90:189.

164. Baumann, B., Danon, L., Weitz, R., Blumensohn, R., Schonfeld, T. and Nitzan, M. Unilateral hydrocephalus due to obstruction of the foramen of Monro: another complication of intrauterine mumps infection? Eur. J. Pediatr., 1982, 139:158-159.

165. Bayatpour, M., Logan, J., Perry, H.P. and Finley, M. Horner's syndrome due to congenital cytomegalovirus infection. South Med. J., 1982, 75:220-221.

166. Beasley, R.P., Hwang, L.Y., Lee, G.C., Lan, C.C., Roan, C.H., Huang, F.Y. and Chen, C.L. Prevention of perinatally transmitted hepatitis B virus infections with hepatitis B virus infections with hepatitis B immune globulin and hepatitis B vaccine. Lancet, 1983, 2:1099-1102.

167. Beasley, R.P., Hwang, L.Y., Lin, C.C., Stevens, C.E., Wang, K.Y., Sun, T.S., Hsieh, F.J. and Szmuness, W. Hepatitis B immune globulin (HBIG) efficacy in the interruption of perinatal transmission of hepatitis B virus carrier state. Initial report of a randomised double-blind placebo-controlled trial. Lancet, 1981, 2:388-393.

168. Beasley, R.P., Shiao, I.S., Wu, T.C. and Hwang, L.Y. Hepatoma in an HBsAg carrier -- seven years after perinatal infection. J. Pediatr., 1982, 101:83-84.

169. Beck, I. and Clayton, J.K. Vaginal carcinoma arising in vaginal condylomata. Case report. Br. J. Obstet. Gynaecol., 1984, 91:503-505.

170. Becroft, D.M. Prenatal cytomegalovirus infection and muscular deficiency (eventration) of the diaphragm. J. Pediatr., 1979, 94:74-75.

171. Becroft, D.M. Prenatal cytomegalovirus infection: epidemiology, pathology and pathogenesis. Perspect. Pediatr. Pathol., 1980, 6:203-241.

172. Beecham, J.B., Braun, T.E., Jr., Clapp, J.F. III and Lucey, J.F. Intrauterine diagnosis of fetal liver dysfunction. Obstet. Gynecol., 1973, 41:556-562.

173. Beers, B.B., Flowers, F.P., Sherertz, E.F. and Selden, S. Dystrophic calcinosis cutis secondary to intrauterine herpes simplex. Pediatr. Dermatol., 1986, 3:208-211.

174. Behbehani, A.M. Virus morbidum foetus. Congenital malformations and neonatal infections. J. Kans. Med. Soc., 1978, 79:653-655.

175. Behlmer, S.D. and Anderson, P.C. Herpes simplex infections complicating parturition. Int. J. Dermatol., 1981, 20:242-248.

176. Belak, S. and Rusvai, M. In utero adenoviral infection of sheep. Vet. Microbiol., 1986, 12:87-91.

177. Belcourt, R.J., Wong, F.C. and Walcroft, M.J. Growth of rubella virus in rabbit foetal tissues and cell cultures. Canad. J. Pub. Health, 1965, 56:253.

178. Belfrage, S. and Dahlquist, E. Infectious mononucleosis. Age, civil state and pregnancy -- an epidemiological study. Scand. J. Infect. Dis., 1969, 1:57-60.

179. Bell, A.M., Kingsley, S.R., Fiddian, A.P. and Brigden, W.D. Sexually transmitted diseases in pregnancy. Br. Med. J., 1984, 288:1456.

180. Bell, E.J., Ross, C.A. and Grist, N.R. ECHO 9 infection in pregnant women with suspected rubella. J. Clin. Pathol., 1975, 28:267-269.

181. Bellanti, J.A. German measles in pregnancy. G.P., 1968, 38:119.

182. Bellanti, J.A., Artenstein, M.S., Olson, L.C., Buescher, E.L., Luhrs, C.E. and Milstead, K.L. Congenital rubella. Clinicopathologic, virologic, and immunologic studies. Am. J. Dis. Child, 1965, 110:464-472.

183. Belman, A.L., Novick, B., Ultmann, M.H., Spiro, A., Rubinstein, A., Horoupian, D.S. and Cohen, H. Neurological complications in children with acquired immune deficiency syndrome. Annals of Neurology, 1984, 16:414.

184. Benirschke, K. Routes and types of infection in the fetus and newborn. Amer. J. Dis. Child, 1960, 99:714.

185. Benirschke, K., Mendoza, G.R. and Bazeley, P.L. Placental and fetal manifestations of cytomegalovirus infection. Virchows Arch., 1974, 16:121-139.

186. Benjamin, B., Brickman, H.F. and Neaga, A. Congenital toxoplasmosis in twins. Can. Med. Assoc. J., 1958, 80:639-643.

187. Bennett, N.M., Forbes, J.A., Lucas, C.R. and Kucers, A. Infective hepatitis and pregnancy: analysis of liver function test results. Med. J. Aust., 1967, 2:974-976.

188. Bennett, N.M., Forbes, J.A., Lucas, C.R. and Kucers, A. Liver disease in pregnancy. Med. J. Aust., 1968, 1:608-609.

189. Bennett, N.M., Lehmann, N.I. and Speed, B.R. Viral hepatitis and intrahepatic cholestasis of pregnancy. Aust. N.Z. J. Med., 1979, 9:54-57.

190. Bennette, J.G., Bush, P.V. and Steele, R.D. Characteristics of a newborn runt disease induced by neonatal infection with an oncolytic strain of reovirus type 3 (REO3MH). II. Immunological aspects of the disease in mice. Br. J. Exp. Pathol., 1967, 48:267-284.

191. Benson, L. and Hotchin, J. Antibody formation in persistent tolerant infection with lymphocytic choriomeningitis virus. Nature, 1969, 222:1045-1047.

192. Benyeda, J., Meszaros, J. and Reibling, J. Foetal disease caused by swine enteroviruses. (Occurrences of "Smedi" disease in Hungary). Acta. Vet. Acad. Sci. Hung., 1973, 23:25-35.

193. Berenberg, W. and Nankervis, G. Long-term follow-up of cytomegalic inclusion disease of infancy. Pediatrics, 1970, 46:403.

194. Bergamini, F. and Ferroni, P. Transplacental acquisition of cytomegalovirus antibody. Observations with two different CMV strains. Boll. Ist. Sieroter. Milan., 1977, 56:15-21.

195. Berger, S.A., Weinberg, M., Treves, T., Sorkin, P., Geller, E., Yedwab, G., Tomer, A., Rabey, M. and Michaeli, D. Herpes encephalitis during pregnancy: failure of acyclovir and adenine arabinoside to prevent neonatal herpes. Isr. J. Med. Sci., 1986, 22:41-44.

196. Bergman, A., Bhatia, N.N. and Broen, E.M. Cryotherapy for treatment of genital condylomata during pregnancy. J. Reprod. Med., 1984, 29:432-435.

197. Bergs, V.V. Rat virus - mediated suppression of leukemia induction by Moloney virus in rats. Cancer Res., 1969, 29:1669-1672.

198. Berkovich, S. and Ressel, M. Neonatal herpes keratitis. J. Pediatr., 1966, 69:652-653.

199. Berkovich, S. and Smithwick, E.M. Transplacental infection due to ECHO virus type 22. J. Pediatr., 1968, 72:94-96.

200. Berman, L.D., Cotes, P.M. and Simons, P.J Tumors and immunological response in rhesus monkeys born after in utero innoculation with Rous sarcoma virus. J. Natl. Cancer Inst., 1967, 39:119-134.

201. Bernstein, D.I. and Ogra, P.L. Fetomaternal apects of immunization with RA27/3 live attenuated rubella virus vaccine during pregnancy. J. Pediatr., 1980, 97:467-470.

202. Berrebi, A., Schattner, A. and Mogilner, B.M. Disseminated Burkitt's lymphoma during pregnancy. Acta. Haematol. (Basel)., 1983, 70:139-140.

203. Berry, P.J. and Nagington, J. Fatal infection with ECHO virus 11. Arch. Dis. Child, 1982, 57:22-29.

204. Best, J.M., Harcourt, G. and Banatvala, J.E. Congenital rubella affecting an infant whose mother had rubella antibodies before conception. Br. Med. J., 1981, 282:1235.

205. Beswick, R.C., Warner, R. and Warkany, J. Congenital anomalies following maternal rubella. Am. J. Dis. Child, 1949, 78:334-348.

206. Bettoli, E.J. Herpes: facts and fallacies. J. Pract. Nurs., 1982, 32:17-21.

207. Betts, R.F. Syndromes of cytomegalovirus infection. Adv. Intern. Med., 1980, 26:447-466.

208. Beverley, J.K.A. Congenital transmission of toxoplasmosis through successive generations of mice. Nature, 1959, 183:1348-1349.

209. Bhalerao, V.R., Desi, V.P. and Pai, D.N. Viral hepatitis in pregnancy. Indian J. Public Health, 1974, 18:165-170.

210. Bharija, S.C., Kanwar, A.J. and Singh, G. Ophthalmic zoster in infancy. A case report. Indian J. Dermatol., 1983, 28:173-174.

211. Bhonde, R.R. and Wagh, U.V. Susceptibility of human placenta to Japanese encephalitis virus in vitro. Indian J. Med. Res., 1985, 82:371-373.

212. Bia, F.J., Griffith, B.P., Tarsio, M. and Hsiung, G.D. Vaccination for the prevention of maternal and fetal infection with guinea pig cytomegalovirus. J. Infect. Dis., 1980, 142:732-738.

213. Bia, F.J., Miller, S.A. and Davidson, K.H. The guinea pig cytomegalovirus model of congenital human cytomegalovirus infection. Birth Defects, 1984, 20:233-241.

214. Bia, F.J., Miller, S.A., Lucia, H.L., Griffith, B.P. and Tarsio, M. and Hsiung, G.D. Vaccination against transplacental cytomegalovirus transmission: vaccine reactivation and efficacy in guinea pigs. J. Infect. Dis., 1984, 149:355-362.

215. Biano, S., Cochran, W., Herrmann, K.L., Hall, A.D. and Chang, T.W. Rubella reinfection during pregnancy. A case of mistaken diagnosis of congenital rubella. Am. J. Dis. Child, 1975, 129:1353-1356.

216. Bielaaski, A. and Raeside, J.I. Plasma concentrations of steroid hormones in sows infected experimentally with Leptospira pomona or porcine enterovirus strain T1 in late gestation. Res. Vet. Sci., 1977, 22:28-34.

217. Biggar, R.J., Henle, W., Fleisher, G., Bocker, J., Lennette, E.T. and Henle, G. Primary Epstein-Barr virus infections in African infants. I. Decline of maternal antibodies and time of infection. Int. J. Cancer, 1978, 22:239-243.

218. Binkhorst, C.D. Toxoplasmosis: a clinical, serological study with special reference to eye manifestations. Leiden, H.E. Stenfert Kroese, 1948, 163.

219. Binkin, N.J. and Alexander, E.R. Neonatal herpes: how can it be prevented? J.A.M.A., 1983, 250:3094-3095.

220. Binkin, N.J., Koplan, J.P. and Cates, W., Jr. Preventing neonatal herpes. The value of weekly viral cultures in pregnant women with recurrent genital herpes. J.A.M.A., 1984, 251:2816-2821.

221. Birdsong, W.M. and Harris, R.E. Rubella susceptibility and syphilis in Asian pregnancies. Am. J. Obstet. Gynecol., 1982, 142:705-706.

222. Birnbaum, G., Lynch, J.I., Margileth, A.M., Lonergan, W.M. and Sever, J.L. Cytomegalovirus infections in newborn infants. J. Pediatr., 1969, 75:789-795.

223. Bistner, S.I., Rubin, L.F. and Saunders, L.Z. The ocular lesions of bovine viral diarrhea - mucosal disease. Pathol. Vet., 1970, 275-286.

224. Bithell, J.F., Draper, G.J. and Gorbach, P.D. Association between malignant disease in children and maternal virus infections during pregnancy. Br. J. Prev. Soc. Med., 1973, 27:68.

225. Bithell, J.F., Draper, G.J. and Gorbach, P.D. Association between malignant disease in children and maternal virus infections. Br. Med. J., 1973, 1:706-708.

226. Bittle, J.L. Comments on biologic requirements and control of equine rhinopneumonitis vaccine. J. Am. Vet. Med. Assoc., 1969, 155:312-314.

227. Bittle, J.L. Newer knowledge on the use of measles vaccine for canine distemper prophylaxis. J. Am. Vet. Med. Assoc., 1970, 156:1752-1755.

228. Bittle, J.L. and Peckham, J.C. Genital infection induced by feline rhinotracheitis virus and effects on newborn kittens. J. Am. Vet. Med. Assoc., 1971, 158:P Suppl 2:97-98.

229. Black, F.L., Berman, L.L., Borgono, J.M., Capper, R.A., Carvalho, A.A., Collins, C., Glover, O., Hijazi, Z., Jacobson, D.L., Lee, Y.L. et al. Geographic variation in infant loss of maternal measles antibody and in prevalence of rubella antibody. Am. J. Epidemiol., 1986, 124:442-452.

230. Black, N. Rubella vaccination. Br. Med. J., 1979, 2:1368.

231. Black, V.H. Virus particles in primordial germ cells of fetal guinea pigs. J. Natl. Cancer Inst., 1974, 52:545-551.

232. Blanc, W.A. Pathology of the placenta and cord in some viral infections. Major Probl. Clin. Pediatr., 1978, 17:237-258.

233. Blanc, W.A. Pathology of the placenta and cord in ascending and in haematogenous infection. Ciba. Found. Symp., 1979, 77:17-38.

234. Blanc, W.A. Pathology of the placenta, membranes, and umbilical cord in bacterial, fungal, and viral infections in man. Monogr. Pathol., 1981, 22:67-132.

235. Bland, P.B. Influenza and its relation to pregnancy and labor. Am. J. Obstet. Gynecol., 1919, 79:184.

236. Blaskovic, D., Rathova, V., Kociskova, D., Kaplan, M.M., Jamrichova, O. and Sadecky, E. Experimental infection of weanling pigs with A-swine influenza virus. 3. Immunity in piglets farrowed by antibody - bearing dams experimentally infected a year earlier. Bull. W.H.O., 1970, 42:771-777.

237. Blattner, R.J. The role of viruses in congenital defects. Am. J. Dis. Child, 1974, 128:781-786.

238. Blattner, R.J. and Heys, F.M. Role of viruses in the etiology of congenital malformations. Prog. Med. Virol., 1961, 3:311.

239. Blattner, R.J., Williamson, A.P. and Heys, F.M. Role of viruses in the etiology of congenital malformations. Prog. Med. Virol., 1973, 15:1-41.

240. Block, B.S. and Goodner, D.M. False - positive amniotic fluid cytology in a parturient with active gential herpes infection at term. Obstet. Gynecol., 1979, 54:658-660.

241. Bloomfield, R., Suarez, J.R., Roque, Z.S. and Tantakasem, P. Herpes simplex in gynecology and obstetrics. J. Natl. Med. Assoc., 1979, 71:161-164.

242. Blot, W.J., Draper, G., Kinlen, L. and Wilson, M.K. Childhood cancer in relation to prenatal exposure to chickenpox. Br. J. Cancer, 1980, 42:342-344.

243. Blumberg, B.S., Sutnick, A.I., London, W.T. and Millman, I. Australia antigen and hepatitis: a comprehensive review. Crit. Rev. Clin. Lab. Sci., 1971, 2:473-528.

244. Bobo, C.B., Antine, B. and Manor, J.P. Neonatal herpes simplex infection limited to the cornea. Arch. Ophthalmol., 1970, 84:697-698.

245. Boehm, F.H., Estes, W., Wright, P.F. and Growdon, J.F., Jr. Management of genital herpes simplex virus infection occurring during pregnancy. Am. J. Obstet. Gynecol., 1981, 141:735-740.

246. Boehme, T.L. Hepatitis B. The nurse - midwife's role in management and prevention. J. Nurse - Midwife, 1985, 30:79-87.

247. Bohl, E.H., Gupta, R.K., McCloskey, L.W. and Saif, L. Immunology of transmissible gastroenteritis. J. Am. Vet. Med. Assoc., 1972, 160:543-549.

248. Bolin, C.A., Bolin, S.R., Kluge, J.P. and Mengeling, W.L. Pathologic effects of intrauterine deposition of pseudorabies virus on the reproductive tract of swine in early pregnancy. Am. J. Vet. Res., 1985, 46:1039-1042.

249. Bolin, S.R., Runnels, L.J., Sawyer, C.A. and Gustafson, D.P. Experimental transmission of pseudorabies virus in swine by embryo transfer. Am. J. Vet. Res., 1982, 43:278-280.

250. Bolognese, R.J. and Corson, S.L. Rubella vaccination. A critical review. Obstet. Gynecol., 1973, 42:851-855.

251. Bolognese, R.J., Corson, S.L., Fuccillo, D.A., Sever, J.L. and Traub, R. Evaluation of possible transplacental infection with rubella vaccination during pregnancy. Am. J. Obstet. Gynecol., 1973, 117:939-941.

252. Bolognese, R.J., Corson, S.L., Fuccillo, D.A. and Traub, R. The susceptibility of the postpartum and postabortal cervix and uterine cavity to infection with attenuated rubella virus. Am. J. Obstet. Gynecol., 1976, 125:525-527.

253. Bolognese, R.J., Corson, S.L., Fuccillo, D.A., Traub, R., Moder, F. and Sever, J.L. Herpesvirus hominis type II infections in asymptomatic pregnant women. Obstet. Gynecol., 1976, 48:507-510.

254. Bolognese, R.J., Corson, S.L., Sever, J.L., Fuccillo, D.A., Lakoff, K.M. and Klein, J. Rubella vaccination during pregnancy. Am. J. Obstet. Gynecol., 1972, 112:903-907.

255. Bolton, W.D., Durrell, W.B., Wadsworth, J.R. and Murray, R.W. A survey of abortions in Vermont dairy cattle. J. Am. Vet. Med. Assoc., 1969, 155:500-503.

256. Bonar, B.E. and Pearsall, C.J. Herpes zoster in the new-born. Amer. J. Dis. Child, 1932, 44:398.

257. Boniuk, M. Glaucoma in the congenital rubella syndrome. Int. Ophthalmol. Clin., 1972, 12:121-136.

258. Boniuk, V. Systemic and ocular manifestations of the rubella syndrome. Int. Ophthalmol. Clin., 1972, 12:67-76.

259. Booth, P.B. and MacGregor, A. The incidence of cold auto - haemagglutinins in Melanesian children and adults. Vox. Sang., 1966, 11:720-723.

260. Bordley, J.E., Brookhouser, P.E., Hardy, J. and Hardy, W.G. Prenatal rubella. Acta. Otolaryngol. (Stockh)., 1968, 66:1-9.

261. Bordley, J.E. and Hardy, J.M. Laboratory and clinical observations on prenatal rubella. Ann. Otol. Rhinol. Laryngol., 1969, 78:917-928.

262. Borgen, H.C. and Ludwig, H. Equine herpesvirus 1: biological and biophysical comparison of two viruses from different clinical entities. Intervirology, 1974, 4:189-198.

263. Borhanmanesh, F., Haghighi, P., Hekmat, K., Rezaizadeh, K. and Ghavami, A.G. Viral hepatitis during pregnancy. Severity and effect on gestation. Gastroenterology, 1973, 64:304-312.

264. Borneman, K. Infectious mononucleosis in third trimester of pregnancy. J. Fam. Pract., 1983, 16:1171-1179.

265. Bornkamm, G.W., Kaduk, B., Kachel, G., Schneider, U., Fresen, K.O., Schwanitz, G. and Hermanek, P. Epstein - Barr virus - postive Burkitt's lymphoma in a German woman during pregnancy. Blut., 1980, 40:167-177.

266. Bortolotti, F., Cadrobbi, P., Bertaggia, A., Rude, L., Alberti, A. and Realdi, G. A 7 year survey of acute hepatitis type B. Arch. Dis. Child, 1983, 58:993-996.

267. Borton, T.E. and Stark, E.W. Audiological findings in hearing loss secondary to maternal rubella. Pediatrics, 1970, 45:225-229.

268. Borton, T.E. and Stark, E.W. More on hearing loss secondary to maternal rubella. J. Speech Hear Disord., 1972, 37:138-139.

269. Borzyskowski, M., Harris, R.F. and Jones, R.W. The congenital varicella syndrome. Eur. J. Pediatr., 1981, 137:335-338.

270. Botha, J.F., Ritchie, M.J., Dusheiko, G.M., Mouton, H.W. and Kew, M.C. Hepatitis B virus carrier state in black children in Ovamboland: role of perinatal and horizontal infection. Lancet, 1984, 1:1210-1212.

271. Bott, L.M. and Eizenberg, D.H. Congenital rubella after successful vaccination. Med. J. Aust., 1982, 1:514-515.

272. Boucher, M. and Yonekura, M.L. Perinatal listeriosis early-onset correlation of antenatal manifestations and neonatal outcome. Obstet. Gynecol., 1986, 68:593-597.

273. Boue, A. and Boue, J.G. Effects of rubella virus infection on the division of human cells. Amer. J. Dis. Child, 1969, 118:45.

274. Boue, A., Nicolas, A. and Montagnon, B. Reinfection with rubella in pregnant women. Lancet, 1971, 1:1251-1253.

275. Boue, A., Papiernick-Berkhauer, E., Levy-Thierry, S. Attenuated rubella virus vaccine in women. Clinical trials during the postpartum period. Am. J. Dis. Child, 1969, 118:230-233.

276. Bourne, F.J., Curtis, J., Johnson, R.H. and Collings, D.F. Antibody formation in porcine fetuses. Res. Vet. Sci., 1974, 16:223-227.

277. Bowen, R.A., Howard, T.H., Elsden, R.P. and Seidel, G.E. Bluetongue virus and embryo transfer in cattle. Prog. Clin. Biol. Res., 1985, 178:85-89.

278. Bowers, V.M. Jr. and Danforth, D.N. The significance of poliomyelitis during pregnancy - an analysis of the literature and presentation of twenty-four new cases. Amer. J. Obst. Gyn., 1953, 65:34.

279. Bowes, W.A., Jr., Gibson, J.L., Leibovitz, A. and Palin, W.J. Rubella antibody screening in a prenatal clinic using the indirect fluorescent method. Obstet. Gynecol., 1970, 35:7-11.

280. Boxaca, M.C., Gomez, M.M., Berria, M.I. and Iacono, R.F. Transplacental infection in guinea pigs inoculated with an attenuated strain of Junin virus. Intervirology, 1984, 21:178-180.

281. Boxall, E.H. Vertical transmission of hepatitis-B surface antigen. Biomedicine, 1982, 26:12-15.

282. Boxall, E.H. Vertical transmission of hepatitis B. Clin. Exp. Obstet. Gynecol., 1982, 10:41-48.

283. Boxall, E.H. Hepatitis problems in the pregnant and nursing mother and the newborn. Midwives Chron., 1983, 96:226-229.

284. Boxall, E.H. and Davies, H. Dane particles in cord blood. Lancet, 1974, 2:1513.

285. Boxall, E.H. and Flewett, T.H. Vertical transmission of hepatitis-B surface antigen. Lancet, 1975, 2:1211.

286. Boxall, E.H., Flewett, T.H., Derso, A. and Tarlow, M.J. Specific immunoglobulin for babies born to HBsAg carriers. Lancet, 1980, 1:419-420.

287. Boxall, E.H. and Tarlow, M.J. Hepatitis B vaccine in the prevention of perinatally transmitted hepatitis B virus infections: initial report of a study in the West Midlands of England. J. Med. Virol., 1986, 18:255-260.

288. Boyd, J.F. Mini-necropsy. Lancet, 1981, 2:1235.

289. Bradley, J.S., Yeager, A.S., Dyson, D.C., Hensleigh, P.A. and Medearis, A.L. Neutralization of herpes simplex virus by antibody in amniotic fluid. Obstet. Gynecol., 1982, 60:318-321.

290. Brady, W.K. and Purdon, A., Jr. Intrauterine fetal demise associated with enterovirus infection. South Med. J., 1986, 79:770-772.

291. Brand, N., Legum, S., Saunders, J. and Fogel, A. Congenital rubella in Israel following the 1978-79 rubella epidemic. Isr. J. Med. Sci., 1983, 19:925-928.

292. Brandling-Bennett, A.D., Wyll, S.A. and Grand, M.G. Current status of rubella in the United States. J. Infect. Dis., 1972, 125:327-329.

293. Braun, R.K., Osburn, B.I. and Kendrick, J.W. Immunologic response of bovine fetus to bovine viral diarrhea virus. Am. J. Vet. Res., 1973, 34:1127-1132.

294. Brazin, S.A., Simkovich, J.W. and Johnson, W.T. Herpes zoster during pregnancy. Obstet. Gynecol., 1979, 53:175-181.

295. Brendel, W. and Ring, J. Immunosuppressive therapy with anti-lymphocytic globulin and tumor formation. In: Gotz, H., Bucherl, E.S., ed. Applied tumor immunology. Berlin, de Gruyter, 1975, 333-347.

296. Britton, W.J., Parsons, C., Gallagher, N.D., Cossart, Y. and Burnett, L. Risk factors associated with hepatitis B infection in antenatal patients. Aust. N.Z. J. Med., 1985, 15:641-644.

297. Broadbent, E., Ajina, N. and Hurley, R. Susceptibility to rubella in a pregnant population after the introduction of vaccination. J. Clin. Pathol., 1980, 33:24-27.

298. Brook, C.G.D. Evidence for a sensitive period of adipose cell replication in man. Lancet, 1972, 2:624.

299. Brookhouser, P.E. and Bordley, J.E. Congenital rubella deafness. Pathology and pathogenesis. Arch. Otolaryngol., 1973, 98:252-257.

300. Brotherston, J.G., Bannatyne, C.C., Mathieson, A.O. and Nicolson, T.B. Field trials of an inactivated oil - adjuvant vaccine against louping - ill (Arbovirus group B). J. Hyg. (Camb)., 1971, 69:479-489.

301. Brotherston, J.G. and Boyce, J.B. Development of a non-infective protective antigen against louping - ill (arbovirus group B). Laboratory experiments. J. Comp. Pathol., 1970, 80:377-388.

302. Brouwer, R., Groot, I.G. de and Verheij, F.B. Comparison of maternal and cord serum titres for measles and for rubella antibodies. Arch. Gesamte. Virusforsch., 1974, 44:237-242.

303. Browder, J.A. Occipital encephalocele after probable prenatal rubella. Lancet, 1971, 2:383-384.

304. Brown, C.C. and MacLachlan, N.J. Congenital encephalopathy in a calf. Vet. Pathol., 1983, 20:770-773.

305. Brown, D.W., Gardner, S.D., Gibson, P.E. and Field, A.M. BK virus specific IgM responses in cord sera, young children and healthy adults detected by RIA. Arch. Virol., 1984, 82:149-160.

306. Brown, G.C. Recent advances in the viral aetiology of congenital anomalies. Adv. Teratol., 1966, 1:55.

307. Brown, G.C. Consequences of natural exposure to rubella during pregnancy. Proc. Soc. Exp. Biol. Med., 1968, 127:340-344.

308. Brown, G.C. Maternal virus infection and congenital anomalies. Arch. Environ. Health, 1970, 21:362-365.

309. Brown, G.C. and Evans, T.N. Serologic evidence of Coxsackievirus etiology of congenital heart disease. J.A.M.A., 1967, 199:183-187.

310. Brown, G.C. and Karunas, R.S. Relationship of congenital anomalies and maternal infection with selected enteroviruses. Am. J. Epidemiol., 1972, 95:207-217.

311. Brown, R.S. The "expanded" rubella syndrome. Va. Med. Mon., 1967, 94:232-237.

312. Brown, S.G. The devastating effects of congenital rubella. M.C.N., 1979, 4:171-173.

313. Brown, T., Anand, A., Ritchie, L.D., Clewley, J.P. and Reid, T.M. Intrauterine parvovirus infection associated with hydrops fetalis. Lancet, 1984, 2:1033-1034.

314. Brown, T., Hambling, M.H. and Ansari, B.M. Rubella - neutralizing and haemagglutinin - inhibiting antibodies in children of different ages. Br. Med. J., 1969, 4:263-265.

315. Brown, T. and Ritchie, L.D. Infection with parvovirus during pregnancy. Br. Med. J. (Clin. Res.), 1985, 290:559-560.

316. Brown, T.T., Bistner, S.I., DeLahunta, A., Scott, F.W. and McEntee, K. Pathogenetic studies of infection of the bovine fetus with bovine viral diarrhea virus. II. Occular lesions. Vet. Pathol., 1975, 12:394-404.

317. Brown, T.T., DeLahunta, A., Bistner, S.I., Scott, F.W. and McEntee, K. Pathogenetic studies of infection of the bovine fetus with bovine viral diarrhea virus. I. Cerebellar atrophy. Vet. Pathol., 1974, 11:486-505.

318. Brown, T.T., DeLahunta, A., Scott, F.W., Kahrs, R.F., McEntee, K. and Gillespie, J.H. Virus induced congenital anomalies of the bovine fetus. II. Histopathology of cerebellar degeneration (hypoplasia) induced by the virus of bovine viral diarrhea - mucosal disease. Cornell Vet., 1973, 63:561-578.

319. Brown, T.T., Mengeling, W.L., Paul, P.S. and Pirtle, E.C. Porcine fetuses with pulmonary hypoplasia resulting from experimental swine influenza virus infection. Vet. Pathol., 1980, 17:455-468.

320. Brown, T.T., Jr., Mengeling, W.L. and Pirtle, E.E. Failure of swine influenza virus to cause transplacental infection of porcine fetuses. Am. J. Vet. Res., 1982, 43:817-819.

321. Brown, T.T., Jr., Schultz, R.D., Duncan, J.R. and Bistner, S.I. Serological response of the bovine fetus to bovine viral diarrhea virus. Infect. Immun., 1979, 25:93-97.

322. Brown, W.J. and Moore, M.B. Jr. Congenital syphilis in the United States. Clin. Pediat., 1963, 2:220.

323. Brown, W.M., Cowper, H.H. and Hodgman, J.E. Gonococcal ophthalmia among newborn infants at Los Angeles County General Hospital 1957-1963. Public Health Rep., 1966, 81:926-928.

324. Brown, Z.A., Berry, S. and Vontver, L.A. Genital herpes simplex virus infections complicating pregnancy. Natural history and peripartum management. J. Reprod. Med., 1986, 31:420-425.

325. Brown, Z.A., Vontver, L.A., Benedetti, J., Critchlow, C.W., Hickok, D.E., Sells, C.J., Berry, S. and Corey, L. Genital herpes in pregnancy: risk factors associated with recurrences and asymptomatic viral shedding. Am. J. Obstet. Gynecol., 1985, 153:24-30.

326. Brownlie, J., Clarke, M.C. and Howard, C.J. Experimental production of fatal mucosal disease in cattle. Vet. Rec., 1984, 114:535-536.

327. Bruhl, H.H., Bahn, R.C. and Hayles, A.B. Sexual precocity associated with congenital toxoplasmosis. Proc. Mayo Clinic, 1958, 33:682-686.

328. Brunell, P.A. Placental transfer of varicella - zoster antibody. Pediatrics, 1966, 38:1034-1038.

329. Brunell, P.A. Varicella - zoster infections in pregnancy. J.A.M.A., 1967, 199:315-317.

330. Brunell, P.A. Interview with Philip A. Brunell: contagion and varicella - zoster virus. Pediatr. Infect. Dis., 1982, 1:304-307.

331. Brunell, P.A. Fetal and neonatal varicella - zoster infections. Semin Perinatol., 1983, 7:47-56.

332. Brunell, P.A. and Kotchmar, G.S., Jr. Zoster in infancy: failure to maintain virus latency following intrauterine infection. J. Pediatr., 1981, 98:71-3.

333. Bryan, C.S., Foster, C.M. and Edwards, W.G. Postpartum cytomegalovirus infection. A hazard of multiple transfusions. Obstet. Gynecol., 1978, 52:68-69.

334. Bryans, J.T. On immunity to disease caused by equine herpesvirus 1. J. Am. Vet. Med. Assoc., 1969, 155:294-300.

335. Bryans, J.T. and Allen, G.P. Application of a chemically inactivated, adjuvanted vaccine to control abortigenic infection of mares by equine herpesvirus I. Dev. Biol. Stand., 1982, 52:493-498.

336. Bryner, J.H., Ritchie, A.E., Booth, G.D. and Foley, J.W. Lytic activity of vibrio phages on strains of Vibrio fetus isolated from man and animals. Appl. Microbiol., 1973, 26:404-409.

337. Buchholz, H.M., Frosner, G.G. and Ziegler, G.B. HBAg carrier state in an infant delivered by caesarean section. Lancet, 1974, 2:343.

338. Budai, J., Toth, M. and Melis, L. Experimental active and passive immunization against rubella. Acta. Paediatr. Acad. Sci. Hung., 1970, 11:209-217.

339. Buffett, R.F. Prevention of 334C murine virus - induced leukemia by transmission of maternal immunity to offspring. Cancer Res., 1974, 34:559-563.

340. Bujko, M., Sulovic, V. and Dotlic, R. Herpesvirus hominis (HVH) infection in women with preterm labor. J. Perinat. Med., 1986, 14:319-324.

341. Bujko-Klisura, M., Sulovic, V., Ovcaric, A., Marinkovic-Zivanovic, V., Markovic, L. and Jovanovic, T. Rapid diagnosis of herpesvirus hominis infection of cervix and vagina before delivery. Clin. Exp. Obstet. Gynecol., 1984, 11:69-71.

342. Bull, M.J. Screening for rubella in women. J. R. Coll. Gen. Pract., 1979, 29:249.

343. Bulova, S.I., Schwartz, E. and Harrer, W.V. Hydrops fetalis and congenital syphilis. Pediatrics, 1972, 49:285.

344. Bundey, S. Adult coeliac disease and neuropathy. Lancet, 1967, 1:851-852.

345. Bunnell, C.E. and Monif, G.R. Interstitial pancreatitis in the congenital rubella syndrome. J. Pediatr., 1972, 80:465-466.

346. Burch, G.E., Sun, S.C., Chu, K.C., Sohol, R.S. and Colcolcough, H.L. Instetstitial and Coxsackievirus B myocarditis in infants and children. J. Am. Med. Ass., 1968, 203:1.

347. Burek, J.D., Roos, R.P. and Narayan, O. Virus - induced abortion. Studies of equine herpesvirus 1 (abortion virus) in hamsters. Lab. Invest., 1975, 33:400-406.

348. Burkart-Jayez, S.F. The effects of congenital rubella on the neonate. J. Neurosurg. Nurs., 1982, 14:173-176.

349. Burke, E.P. and Reagles, K.W. Deaf - blindness and the 1963 - 1965 rubella epidemic: rehabilitation challenge for the 1980's. J. Rehabil., 1979, 45:24-27.

350. Burkinshaw, J., Kirman, B.H. and Sorsby, A. Toxoplasmosis in relation to mental deficiency. Br. Med. J., 1953, 1:702-704.

351. Burns, S.J. Equine infectious anemia: plasma clearance times of passively transferred antibody in foals. J. Am. Vet. Med. Assoc., 1974, 164:64-65.

352. Burns, S.M., Cubie, H., Hargreaves, F.D. and Inglis, J.M. Possible loophole in rubella screening of antenatal patients. Lancet, 1982, 2:1046.

353. Burns, W.A. Rubella and abortion laws. Med. Leg. Bull., 1970, 19:1-7.

354. Burrows, R., Goodridge, D. and Denyer, M.S. Trials of an inactivated equine herpesvirus 1 vaccine: challenge with a subtype 1 virus. Vet. Rec., 1984, 114:369-374.

355. Burrows, R., Mann, J.A., Goodridge, D., Wrathall, A.E. and Done, J.T. Swine vesicular disease -- studies in pregnant sows. Zentralbl. Veterinarmed., 1977, 24:177-182.

356. Burston, G.R. Foetal malformations associated with maternal infective hepatitis. Practitioner, 1966, 197:664-666.

357. Burton, A.E., Robinson, E.T., Harper, W.F., Bell, E.J. and Boyd, J.F. Fetal damage after accidental polio vaccination of an immune mother. J. R. Coll. Gen. Pract., 1984, 34:390-394.

358. Burton, J.F. Frequency of rubella antibody in sera of young pregnant women and its implications. N.Z. Med. J., 1967, 66:378-380.

359. Bustamante-Calvillo, M.E., Caloca-Garcia, M.A. and Ruiz-Gomez, J. Hepatitis A. I. Persistence of passive maternal immunity in infants aged less than six months. Arch. Invest. Med. (Mex.), 1982, 13:69-72.

360. Butler, M. and Goffe, A.P. Isolation of a transmissible agent resembling rubella virus from an apparently healthy human embryo lung cell culture. J. Gen. Virol., 1968, 2:285-290.

361. Butler, N.R., Dudgeon, J.A., Hayes, K., Peckham, C.S. and Wybar, K. Persistence of rubella antibody with and without embryopathy. A follow-up study of children exposed to maternal rubella. Br. Med. J., 1965, 5469:1027-1029.

362. Buxton, D., Reid, H.W., Finlayson, J., Pow, I., Berrie, E. Transmission of a malignant catarrhal fever - like syndrome to sheep: preliminary experiments. Res. Vet. Sci., 1985, 38:22-29.

363. Bwangamoi, O. Pathology of ovine foetus infection with bluetongue virus. Bull. Anim. Health Prod. Afr., 1978, 26:78-97.

# C


364. Cabrey, E.A., Stewart, W.C., Young, S.H., Richardson, G.C. Transmission of hog cholera by pregnant sows. J. Am. Vet. Med. Assoc., 1966, 149:23-30.

365. Came, P.E. and Crowell, R.L. Studies of resistance of fetal mouse tissues in culture to Coxsackie A viruses. Virology, 1964, 23:542.

366. Campbell, A.D. Swine fever. Proc. R. Soc. Med., 1969, 62:824-826.

367. Campbell, C.H. Anaphylaxis in pregnant and mother mice following active and passive sensitization. Cornell Vet., 1965, 55:545-558.

368. Cappel, R., DeCuyper, F. and VanBeers, D. Diagnosis of hepatitis B by Dane particle associated DNA polymerase assay. J. Med. Virol., 1978, 3:77-80.

369. Capponi, M.A. Case of congenital trypanosomiasis at Douala, French Equatorial Africa. Trop. Dis. Bull., 1954, 51:356.

370. Carlstrom, G. Viral infections and birth defects. Acta. Obstet. Gynecol. Scand., 1977, 21-31.

371. Carney, J.F., Caroline, N.L., Nankervis, G.A. and Pomeranz, J.R. Eczema vaccinatum and eczeme herpeticum in Darier disease. Arch. Dermatol., 1973, 107:613-614.

372. Carrington, D., Gilmore, D.H., Whittle, M.J., Aitken, D., Gibson, A.A., Patrick, W.J., Brown, T., Caul, E.O., Field, A.M., Clewley, J.P. et al. Maternal serum alpha-fetoprotein -- a marker of fetal aplastic crisis during intrauterine human parvovirus infection. Lancet, 1987, 1:433-435.

373. Carruthers, D.G. Congenital deaf-mutism as sequela of rubella-like maternal infection during pregnancy. Med. J. Aust., 1945, 1:315-320.

374. Cartwright, S.F., Lucas, M. and Huck, R.A. A small haemaglutinating porcine DNA virus. I. Isolation and properties. J. Comp. Pathol., 1969, 79:371-377.

375. Cartwright, S.F., Lucas, M. and Huck, R.A. A small haemaglutinating porcine DNA virus. II. Biological and serological studies. J. Comp. Pathol., 1971, 81:145-155.

376. Carver, D.H. Comment on rubella controversy. Pediatrics, 1975, 55:4-5.

377. Carver, D.H. and Seto, D.S. Current concepts concerning the hepatitis viruses. Pediatrics, 1973, 51:115-119.

378. Carver, D.H., Seto, D.S.Y., Marcus, P.I. and Rodrigues, L. Rubella virus replication in the brains of suckling mice. J. Virol., 1967, 1:1089.

379. Casaro, A.P., Kendrick, J.W. and Kennedy, P.C. Response of the bovine fetus to bovine viral diarrhea - mucosal disease virus. Am. J. Vet. Res., 1971, 32:1543-1562.

380. Cass, V.M., Daniels, J. and MacCready, R.A. Ladies, put your serum in a bank. N. Engl. J. Med., 1969, 281:448-449.

381. Cassell, G.H. Ureaplasmas of human: with emphasis upon maternal and neonatal infections. Future considerations: maternal and neonatal aspects. Pediatr. Infect. Dis., 1986, 5:S341-344.

382. Castle, W.E. Deafness and rubella: infants in the '60s, adults in the '80s. Introduction. Am. Ann. Deaf., 1980, 125:961-962.

383. Catalano, L.W., Jr., Fuccillo, D.A., Traub, R.G. and Sever, J.L. Isolation of rubella virus from placentas and throat cultures of infants. A prospective study after the 1964-65 epidemic. Obst. Gyn., 1971, 38:6.

384. Catalano, L.W., Jr. and Sever, J.L. The role of viruses as causes of congenital diseases. Annu. Rev. Microbiol., 1971, 25:255-282.

385. Cates, W., Jr. Treatment of rabies exposure during pregnancy. Obstet. Gynecol., 1974, 44:893-896.

386. Cates, W., Jr. and Hattwick, M. Rabies in pregnancy. Obstet. Gynecol., 1975, 46:500.

387. Cederqvist, L.L., Abdel-Latif, N., Meyer, J. and Doctor, L. Fetal and maternal humoral immune response to cytomegalovirus infection. Obstet. Gynecol., 1986, 67:214-216.

388. Cederqvist, L.I., Zervoudakis, J.A., Ewool, L.C., Senterfit, L.B. and Litwin, S.D. Prenatal diagnosis of congenital rubella. Obstet. Gynecol., 1977, 32:592-593.

389. Celermajer, J.M., Varghese, P.J. and Rowe, R.D. Cardiovascular lesions in rubella embryopathy with special emphasis on pulmonary arterial disease. Isr. J. Med. Sci., 1969, 5:568-571.

390. Celko, A., Burianova, B., Fedova, D., Koza, J., Zikmundova, V., Zikmund, V. and Stribrny, J. Transplacental antibodies. Part III: Maternal antibodies against polioviruses, M. parotitidis and M. parainfluenzae. J. Hyg. Epidemiol. Microbiol. Immunol., 1985, 29:193-199.

391. Cesario, T.C. Viral infections that affect the fetus. Postgrad. Med., 1976, 59:66-72.

392. Chagnon, A. and Pavilanis, V. Epidemiological studies on rubella. Canad. Med. Assoc. J., 1970, 102:933.

393. Chakravorty, M.S., Mukherjee, M.K., Das, B.C., Bose, A.K., Choudhuri, A. and Sarkar, J.K. Clinical and laboratory findings in two cases of rubella -- a preliminary report. Bull. Calcutta. Sch. Trop. Med., 1972, 20:22-23.

394. Chamberlain, M.J., Reynolds, A.L. and Yeoman, W.B. Medical memoranda. Toxic effect of podophyllum application in pregnancy. Br. Med. J., 1972, 3:391-392.

395. Chan, M.K. and Moorhead, J.F. Hepatitis B and the dialysis and renal transplantation unit. Nephron., 1981, 27:229-232.

396. Chan, S.H., Tan, K.L., Goh, K.T., Lim, C., Tsakok, M., Oon, C.J. and Ratnam, S.S. Maternal - child hepatitis B virus transmission in Singapore. Int. J. Epidemiol., 1985, 14:173-177.

397. Chandler, S.H., Alexander, E.R. and Holmes, K.K. Epidemiology of cytomegaloviral infection in a heterogeneous population of pregnant women. J. Infect. Dis., 1985, 152:249-256.

398. Chang, C.P., New, A.E., Chiang, H.S., Irving, G.S., Taylor, J.F. and Chen, W.F. A surveillance for Japanese encephalitis (JE) in pigs in southern Taiwan. Chin. J. Microbiol., 1974, 7:6-12.

399. Chang, R.S. and Seto, D.S. Perinatal infection by Epstein - Barr virus. Lancet, 1979, 2:201.

400. Chang, T.H., Moorhead, P.S., Boue, J.G., Plotkin, S.A. and Hoskins, J.M. Chromosome studies of human cells infected in utero and in vitro with rubella virus. Proc. Soc. Exp. Biol. Med., 1966, 122:236-243.

401. Chang, T.W. Herpes simplex virus infection. Int. J. Dermatol., 1983, 22:1-7.

402. Chang, T.W. Rubella reinfection and intrauterine involvement. J. Pediatr., 1974, 84:617-618.

403. Chang, T.W., Moorhead, P.S., Boue, J.G., Plotkin, S.A. and Hoskins, J.M. Chromosome studies of human cells infected with in utero and in vitro rubella virus. Proc. Soc. Exp. Biol. Med., 1966, 112:236-243.

404. Chang, T.W. and OKeefe, P. Cesarean section and genital herpes. N. Engl. J. Med., 296:573.

405. Chanock, R.M., Kaplan, A.Z., Mills, J., Kim, H.W. and Parrott, R.H. Influence of immunological factors in respiratory syncytial virus disease. Arch. Environ. Health, 1970, 21:347-355.

406. Chantler, J.K., Misra, V. and Hudson, J.B. Vertical transmission of murine cytomegalovirus. J. Gen. Virol., 1979, 42:621-625.

407. Chapman, A.L., Weitlauf, H.M. and Bopp, W. Effect of feline leukemia virus on transferred hamster fetuses. J. Natl. Cancer Inst., 1974, 52:583-586.

408. Chattopadhyay, B. and Honeycombe, J. HBsAg carriers in an antenatal clinic and their management in labour. Br. J. Clin. Pract., 1978, 32:285-287.

409. Chaturvedi, U.C., Mathur, A., Chandra, A., Das, S.K., Tandon, H.O., Singh, U.K. Transplacental infection with Japanese encephalitis virus. J. Infect. Dis., 1980, 141:712-715.

410. Chaturvedi, U.C., Tripathi, B.N., Mathur, A., Singh, U.K. and Mehrotra, R.M. Role of rubella in congenital malformations in India. J. Hyg. (Camb.), 1976, 76:33-40.

411. Chaudhary, R.K. Perinatal transmission of hepatitis B virus. Can. Med. Assoc. J., 1983, 128:664-666.

412. Check, W.A. Pregnant women still vulnerable to rubella (news). J.A.M.A., 1981, 245:325-326.

413. Cherry, J.D., Soriano, F. and Jahn, C.L. Search for perinatal viral infection. A prospective, clinical, virologic, and serologic study. Am. J. Dis. Child, 1968, 116:245-250.

414. Chess, S. Autism in children with congenital rubella. J. Autism Child Schizo., 1971, 1:33-47.

415. Chess, S. and Fernandez, P. Neurologic damage and behavior disorder in rubella children. Am. Ann. Deaf., 1980, 125:998-1001.

416. Chess, S., Korn, S.J. and Fernandez, P.B. Psychiatric disorders of children with congenital rubella. New York, Brunner/Mazel, 1971.

417. Chiba, Y., Ogra, P.L. and Nakao, T. Transplacental mumps infection. Am. J. Obstet. Gynecol., 1975, 122:904-905.

418. Chieco-Bianchi, L., Collavo, D., Biasi, G. and Colombatti, A. Prevention of murine sarcoma virus oncogenesis in offspring of immunized female mice. Br. J. Cancer, 1973, 28:238-244.

419. Chin, J. Exposure of expectant mothers to cytomegalovirus infections. West. J. Med., 1983, 139:361.

420. Chin, J. Prevention of chronic hepatitis B virus infection from mothers to infants in the United States. Pediatrics, 1983, 71:289-292.

421. Chin, J. Prevention of hepatitis B in infants born to HBsAg - carrier mothers. West J. Med., 1984, 140:933-934.

422. Chin, J., Ebbin, A.J., Wilson, M.G. and Lennette, E.H. Avoidance of rubella immunization of women during or shortly before pregnancy. J.A.M.A., 1971, 215:632-634.

423. Chin, J., Magoffin, R.L. and Lennette, E.H. The need for routine rubella antibody testing of women. Calif. Med., 1972, 116:9-13.

424. Chin, K.C., Chaudhuri, A.K. and Follett, E.A. Specific immunoglobulin for babies born to HBsAg carriers. Lancet, 1980, 1:420.

425. Chin, K.C., Chaudhuri, A.K. and Follett, E.A. Materno - foetal transmission of HBsAg by asymptomatic carrier mothers in Glasgow. J. Infect., 1981, 3:246-252.

426. Chodos, W.S. Varicella in pregnancy: report of case and review of the literature. J. Am. Osteopath. Assoc., 1982, 81:644-646.

427. Choi, Y.C. and Hsiung, G.D. Cytomegalovirus infection in guinea pigs. II. Transplacental and horizontal transmission. J. Infect. Dis., 1978, 138:197-202.

428. Chong, K.T. and Mims, C.A. Delayed hypersensitivity to murine cytomegalovirus and its depression during pregnancy. Infect. Immun., 1982, 37:54-59.

429. Chow, T.C. and Hsiung, G.D. Neonatal herpes simplex virus infection in guinea pigs. Proc. Soc. Exp. Biol. Med., 1982, 170:459-463.

430. Chowdhury, S.I., Kubin, G. and Ludwig, H. Equine herpes virus type 1 (EHV - 1) induced abortions and paralysis in a Lipizzaner stud: a contribution to the classification of equine herpesviruses. Arch. Virol., 1986, 90:273-288.

431. Christie, J.D., Rakusan, T.A., Martinez, M.A., Lucia, H.L., Rajaraman, S., Edwards, S.B. and Hayden, C.K., Jr. Hydranencephaly caused by congenital infection with herpes simplex virus. Pediatr. Infect. Dis., 1986, 5:473-478.

432. Chu, H.P. and Rizk, J. The effect of maternal immunity, age at vaccination and doses or live vaccines on immune response to Newcastle disease. Dev. Biol. Stand., 1975, 28:451-463.

433. Chuang, T.Y., Perry, H.O., Kurland, L.T. and Ilstrup, D.M. Condyloma acuminatum in Rochester, Minn., 1950-1978. II. Anaplasias and unfavorable outcomes. Arch. Dermatol., 1984, 120:476-483.

434. Cibis, A. and Burde, R.M. Herpes simplex virus-induced congenital cataracts. Arch. Ophthal., 1971, 85:220.

435. Clark, H.F. and Karzon, D.T. Suckling mouse cataract agent (SMCA) in mice. II. Transfer of protection of progeny of SMCA - infected dams. J. Immunol., 1968, 101:782-787.

436. Clark, T.L. and Dillman, R.C. Clinicopathology of equine rhinopneumonitis abortion in central Iowa. Vet. Med. Small. Anim. Clin., 1974, 69:320.

437. Clarke, W.L., Shaver, K.A., Bright, G.M., Rogol, A.D. and Nance, W.E. Autoimmunity in congenital rubella syndrome. J. Pediatr., 1984, 104:370-373.

438. Claypool, J.M. Rubella protection for maternal child health care providers. M.C.N., 1981, 6:53-56.

439. Cloonan, M.J., Hawkes, R.A. and Stevens, L.H. Postnatal decline of maternally acquired rubella antibodies. J. Hyg. (Camb.), 1970, 68:461-468.

440. Cloonan, M.J., Hawkes, R.A. and Stevens, L.H. Postnatal decline of maternally acquired viral antibodies of different specificities. J. Hyg. (Camb.), 1971, 69:435-444.

441. Cockburn, W.C. World aspects of the epidemiology of rubella. Am. J. Dis. Child, 1969, 118:112-122.

442. Coetzer, J.A. and Barnard, B.J. Hydrops amnii in sheep associated with hydraencephaly and arthrogryposis with Wesselsbron disease and Rift Valley fever viruses as aetiological agents. Onderstepoort J. Vet. Res., 1977, 44:119-126.

443. Coetzer, J.A., Theodoridis, A., Herr, S. and Kritzinger, L. Wesselsbron disease: a cause of congenital porencephaly and cerebellar hypoplasia in calves. Onderstepoort J. Vet. Res., 1979, 46:165-169.

444. Coffey, V.P. and Jessop, W.J.E. Maternal influenza and congenital deformities. A prospective study. Lancet, 1959, 2:935.

445. Coggin, J.H., Ambrose, K.R. and Anderson, N.G. Fetal antigen capable of inducing transplantation immunity against SV40 hamster tumor cells. J. Immunol., 1970, 105:524-526.

446. Coggin, J.H., Jr., Ambrose, K.R., Bellomy, B.B. and Anderson, N.G. Tumor immunity in hamsters immunized with fetal tissues. J. Immunol., 107:526-533.

447. Cohen, M.M., Sr. The effect of maternal rubella on dental development. Birth Defects, 1974, 10:25-29.

448. Cohen, S.M., Collins, D.N., Ward, G. and Deibel, R. Transplacental transmission of rubella virus infection in rabbits. Appl. Microbiol., 1971, 21:76-78.

449. Cohen, S.M., Deibel, R., Collins, D.N., Schryver, G.D. and Ducharme, C.P. Effect of rubella vaccine on transplacental rubella virus infection in rabbits. J. Lab. Clin. Med., 1972, 80:179-189.

450. Cohn, A.M., Kos, J.T., II, Taber, L.H. and Adam, E. Recurring laryngeal papilloma Am. J. Otolaryngol., 1981, 2:129-132.

451. Coid, C.R. Comparative aspects of infection in pregnancy. In: Intra-uterine Infections, Ciba Foundation Symposium. (North Holland, Amsterdam: Elsevier, Excerpta Medica), 1973, 117-133.

452. Coid, C.R. Infection in pregnancy and impaired fetal growth. Proc. R. Soc. Med., 1976, 69:365-366.

453. Coid, C.R., Lansdown, A.B.G. and MacFadyen, I.R. Fetal growth retardation and low birthweight following infection in pregnancy. In: C.R. Coid (ed.), Infections and Pregnancy, 1977, 289-305.

454. Coid, C.R. and Ramsden, D.B. Retardation of foetal growth and plasma protein development in foetuses from mice injected with Coxsackie B3 virus. Nature, 1973, 241:460-461.

455. Coid, C.R., Ramsden, D.B. and Healy, M.J. Fetal mouse weights and albumin-alpha 1-fetoprotein ratios after maternal infection with Coxsackievirus B3. Med. Microbiol. Immunol. (Berl.), 1974, 159:285-288.

456. Coid, C.R. and Wardman, G. The effect of para-influenza type 1 (Sendai) virus infection on early pregnancy in the rat. J. Reprod. Fertil., 1971, 24:39-43.

457. Coid, C.R. and Wardman, G. The effect of maternal respiratory disease induced by para-influenza type 1 (Sendai) virus on foetal development and neonatal mortality in the rat. Med. Microbiol. Immunol. (Berl.), 1972, 157:181-185.

458. Coignoul, F.L. and Cheville, N.F. Pathology of maternal genital tract, placenta, and fetus in equine viral arteritis. Vet. Pathol., 1984, 21:333-340.

459. Coleman, D.V., Daniel, R.A., Gardner, S.D., Field, A.M. and Gibson, P.E. Polyoma virus in urine during pregnancy. Lancet, 1977, 2:709-710.

460. Coleman, D.V., Gardner, S.D., Mulholland, C., Fridiksdottir, V., Porter, A.A., Lilford, R. and Valdimarsson, H. Human polyomavirus in pregnancy. A model for the study of defence mechanisms to virus reactivation. Clin. Exp. Immunol., 1983, 53:289-296.

461. Coleman, D.V., Wolfendale, M.R., Daniel, R.A., Dhanjal, N.K., Gardner, S.D., Gibson, P.E. and Field, A.M. A prospective study of human polyomavirus infection in pregnancy. J. Infect. Dis., 1980, 142:1-8.

462. Collie, M.H., Rushton, D.I., Sweet, C., Husseini, R.H. and Smith, H. Ferret foetal infection with influenza virus at early gestation. Br. J. Exp. Pathol., 1982, 63:299-304.

463. Collie, M.H., Sweet, C., Cavanagh, D. and Smith, H. Association of foetal wastage with influenza infection during ferret pregnancy. Br. J. Exp. Pathol., 1978, 59:190-195.

464. Collmann, R.D., Krupinski, J. and Stoller, A. Incidence of infectious hepatitis compared with the incidence of children with Down's syndrome born nine months later to younger and to older mothers. J. Ment. Defic. Res., 1966, 10:266-268.

465. Comas, A.P. and Bentances, R.E. Congenital rubella and acquired hypothyroidism secondary to Hashimoto thyroiditis. J. Pediatr., 1976, 88:1065.

466. Combes, B., Shore, G.M., Cunningham, F.G., Walker, F.B., Shorey, J.W. and Ware, A. Serum gamma-glutamyl transpeptidase activity in viral hepatitis: suppression in pregnancy and by birth control pills. Gastroenterology, 1977, 72:271-274.

467. Conner, G.H. and Carter, G.R. Response of the bovine fetus to reovirus. Vet. Med. Small Anim. Clin., 1975, 70:1463-1464.

468. Connolly, J.H., Nevin, N.C., Simpson, D.M. and ONeill, H.J. Outcome of pregnancy in a rubella outbreak in Northern Ireland 1978-1979. Ulster. Med. J., 1984, 53:65-73.

469. Connon, A.F. Risk of rubella in pregnancy. Med. J. Aust., 1975, 1:199-201.

470. Constantine, D.G. Absence of prenatal infection of bats with rabies virus. J. Wildl. Dis., 1986, 22:249-250.

471. Cooper, L.Z. Rubella in pregnancy. Postgrad. Med., 1969, 46:106-107.

472. Cooper, L.Z. Rubella in the U.S.A., the problems. Scand. J. Infect. Dis., 1972, 6:7-13.

473. Cooper, L.Z. Congenital rubella in the United States. In: Krugman, S., Gershon, A.A. ed. Infections of the fetus and the newborn infant. New York, Liss, 1975, 1-22.

474. Cooper, L.Z., Green, R.H., Krugman, S., Giles, J.P. and Mirick, G.S. Neonatal thrombocytopenic purpura and other manifestations of rubella contracted in utero. Am. J. Dis. Child, 1965, 110:416-427.

475. Cooper, L.Z. and Krugman, S. Diagnosis and management: congenital rubella. Pediatrics, 1968, 37:335-338.

476. Cooper, L.Z. and Krugman, S. Clinical manifestations of postnatal and congenital rubella. Arch. Ophthalmol., 1969, 77:434-439.

477. Cooper, L.Z., Ziring, P.R., Ockerse, A.B., Fedun, B.A., Kiely, B. and Krugman, S. Rubella. Clinical manifestations and management. Am. J. Dis. Child, 1969, 118:18-29.

478. Copelovici, Y. Results obtained in the "Stefan S. Nicolau" Institute of Virology on the role of some viral and rickettsial infections in perinatal pathology. Virologie, 1979, 30:273-281.

479. Coria, M.F. and McClurkin, A.W. Isolation of bovine adenovirus type 1 without an adenovirus - associated virus. Am. J. Vet. Res., 1978, 39:1975-1976.

480. Correa, W.M. A rapid method for the diagnosis of equine virus abortion. Can. J. Comp. Med., 1970, 34:164-166.

481. Correa, W.M. and Nilsson, M.R. Hepatitis due to equine abortion virus. Comparison between the liver histology in human, canine, duckling, and equine viral hepatitis. Can. J. Comp. Med., 1966, 30:112-116.

482. Cossart, Y.E. The outcome of hepatitis B virus infection in pregnancy. Postgrad. Med. J., 1977, 53:610-613.

483. Cossart, Y.E. Transmission of hepatitis B from mother to infant. Med. J. Aust., 1978, 2:550-551.

484. Cossart, Y.E., Hargreaves, F.D. and March, S.P. Australia antigen and the human fetus. Am. J. Dis. Child, 1972, 123:376-378.

485. Costa, S., Barrasso, R., Terzano, P., Zerbini, M., Carpi, C. and Musiani, M. Detection of active Epstein-Barr infection in pregnant women. Eur. J. Clin. Microbiol., 1985, 4:335-336.

486. Cotlier, E. Rubella in animals and experimental ocular aspects of congenital rubella. Int. Ophthalmol. Clin., 1972, 12:137-146.

487. Cotlier, E., Fox, J., Bohigian, G., Beaty, G. and DuPree, A. Pathogenic effects of rubella virus on embryos and newborn rats. Nature (Lond.), 1968, 217:38.

488. Counter, C.E. and Korn, B.J. Herpes zoster in the newborn associated with congenital blindness: report of a case. Arch. Pediat., 1950, 67:397.

489. Couvreur, J. and Desmonts, G. Congenital and maternal toxoplasmosis. A review of 300 congenital cases. Dev. Med. Child Neurol., 1962, 4:519.

490. Cowan, M.J., Hellman, D., Chudwin, D., Wara, D.W., Chang, R.S. and Ammann, A.J. Maternal transmission of acquired immune deficiency syndrome. Pediatrics, 1984, 73:382-386.

491. Cox, S.M., Phillips, L.E., DePaolo, H.D. and Faro, S. Treatment of disseminated herpes simplex virus in pregnancy with parenteral acyclovir. A case report. J. Reprod. Med., 1986, 31:1005-1007.

492. Cradock-Watson, J.E., Ridehalgh, M.K., Anderson, M.J. and Pattison, J.R. Outcome of asymptomatic infection with rubella virus during pregnancy. J. Hyg. (Lond.), 1981, 87:147-154.

493. Cradock-Watson, J.E., Ridehalgh, M.K., Anderson, M.J., Pattison, J.R. and Kangro, H.O. Fetal infection resulting from maternal rubella after the first trimester of pregnancy. J. Hyg.(Lond.), 1980, 85:381-391.

494. Cranage, M.P., Gardner, P.S. and McIntosh, K. In vitro cell-dependent lysis of respiratory syncytial virus-infected cells mediated by antibody from local respiratory secretions. Clin. Exp. Immunol., 1981, 43:28-35.

495. Crane, C.S. Infectious bovine rhinotracheitis abortion and its relationship to nutrition in California beef cattle. J. Am. Vet. Med. Assoc., 1965, 147:1308-1309.

496. Cremin, B.J. and Fisher, R.M. The lesions of congenital syphilis. Brit. J. Radiol., 1970, 43:333.

497. Crenshaw, G.L. and McKercher, D.G. Virus abortion in California Proc. U.S. Anim. Health Assoc., 1967, 71:128-132.

498. Crispens, C.G., Jr. Lactate dehydrogenase virus and mouse embryos. Nature, 1967, 214:819.

499. Cristea, A., Copelovici, Y., Cajal, N., Strulovici, D., Teleguta, L., Niculescu, R., Tanase, M., Teodoru, G., Major, E., Anichitoaie, E., et al. Evaluation of the prematurity risk on the ground of severity scores reflecting the action of different factors, maternal virus infection included. Virologie, 1984, 35:21-30.

500. Cristea, A., Copelovici, Y., Cajal, N., Teleguta, L., Niculescu, R., Dinca, A., Grosser, F., Teodoru, G., Major, E., Pavel, A., Frumm, I. and Tirsia, E. Statistical studies on the epidemiology of human congenital malformations. Implications of maternal virus infections. Virologie, 1982, 33:101-115.

501. Crittenden, L.B. The epidemiology of avian lymphoid leukosis. Cancer Res., 1976, 36:570-573.

502. Cropper, M., Dunne, H.W., Leman, A.D., Starkey, A.L. and Hoefling, D.C. Prevalence of antibodies to porcine enteroviruses and porcine parvovirus in body fluids of fetal pigs from small vs. large litters. J. Am. Vet. Med. Assoc., 1976, 168:233-235.

503. Crovari, P., Caffarena, G., Coppola, R., DiPietro, P., Gasparini, R., Gelli, G.P., Icardi, G.C. and Pantarotto, M.F. Passive-active immunoprophylaxis in newborns to HBsAg carrier mothers liguria -- I. Boll. Ist. Sieroter. Milan, 1984, 63:290-298.

504. Csiza, C.K. Comments on serologic procedures in the study of feline panleukopenia. J. Am. Vet. Med. Assoc., 1971, 158:884-887.

505. Csiza, C.K., Scott, F.W., Gillespie, J.H. and DeLahunta, A. Feline viruses. XIV. Transplacental infections in spontaneous panleukopenia of cats. Cornell Vet., 1971, 61:423-439.

506. Curnen, M.G., Varma, A.A., Christine, B.W. and Turgeon, L.R. Childhood leukemia and maternal infectious diseases during pregnancy. J. Natl. Cancer Inst., 1974, 53:943-947.

507. Cutler, R.S., Molitor, T.W., Leman, A.D. and Sauber, T.E. Effect of porcine parvovirus serostatus on the reproductive performance of mated gilts in an infected herd. Am. J. Vet. Res., 1982, 43:935-937.

508. Cutler, R.S., Molitor, T.W., Leman, A.D. and Werdin, R.E. Farm studies of porcine parvovirus infection. J. Am. Vet. Med. Assoc., 1983, 182:592-594.

509. Cutlip, R.C., Lehmkuhl, H.D., Brogden, K.A. and Bolin, S.R. Mastitis associated with ovine progressive pneumonia virus infection in sheep. Am. J. Vet. Res., 1985, 46:326-328.

510. Cutlip, R.C., Lehmkuhl, H.D. and Jackson, T.A. Intrauterine transmission of ovine progressive pneumonia virus. Am. J. Vet. Res., 1981, 42:1795-1797.

511. Cutlip, R.C., Lehmkuhl, H.D., Whipp, S.C. and McClurkin, A.W. Effects on ovine fetuses of exposure to ovine progressive pneumonia virus. Am. J. Vet. Res., 1982, 43:82-85.

512. Cutlip, R.C. and Mengeling, W.L. Pathogenesis of in utero infection: experimental infection of eight- and ten-week-old porcine fetuses with porcine parvovirus. Am. J. Vet. Res., 1975, 36:1751-1754.

513. Cutter, B.G. Neonatal varicellas. Med. J. Aust., 1968, 1:411-412.

514. Czeizel, A. Coxsackievirus and congenital malformation. J.A.M.A., 1967, 201:142.

# D


515. Daffos, F., Forestier, F., Grangeot-Keros, L., Capella-Pavlovsky, M., Lebon, P., Chartier, M. and Pillot, J. Prenatal diagnosis of congenital rubella. Lancet, 1984, 2:1-3.

516. Dale, R.A., Jr. Maternal and neonatal herpes infection. J. Miss. State Med. Assoc., 1976, 17:261-265.

517. Daling, J.R. and Wolf, M.E. The role of decision and cost analyses in the treatment of pregnant women with recurrent genital herpes. J.A.M.A., 1984, 251:2828-2829.

518. Dall, V., Jensen, N.H. and Hansen, U. Diagnosis of pre- and postnatal rubella by demonstration of specific IgM-class antibodies by a microplate immunofluorescence test. Med. Microbiol. Immunol. (Berl.), 1980, 168:73-80.

519. Dalldorf, G. and Gifford, R. Susceptibility of gravid mice to Coxsackievirus infection. J. Exp. Med., 1954, 99:21.

520. Dalldorf, G. and Sickles, G.M. An unidentified filterable agent isolated from the faeces of children with paralysis. Science, 1948, 108:61.

521. Dana, J.B. Coxsackie virus infections and heart disease. J. Maine Med. Assoc., 1968, 59:147.

522. Darby, C.W. and Morgan-Capner, P. Rash in early pregnancy. Br. Med. J., 1983, 287:202.

523. Dardi, L.E., Ariano, L., Ariano, M.C. and Gould, V.E. Arias-Stella reaction with prominent nuclear pseudoinclusions simulating herpetic endometritis. Diagn. Gynecol. Obstet., 1982, 4:127-132.

524. Davenport, F.M. Pathogenesis of influenza. Bacteriol., 1961, 25:294.

525. David, T.J. and Williams, M.L. Herpes zoster in infancy. Scand. J. Infect. Dis., 1979, 11:185-186.

526. Davies, F.G. and Karstad, L. Experimental infection of the African buffalo with the virus of Rift Valley fever. Trop. Anim. Health Prod., 1981, 13:185-188.

527. Davies, F.G., Soi, R.K. and Wariru, B.N. Abortion in sheep caused by Thogoto virus. Vet. Rec., 1984, 115:654.

528. Davies, P.A. Maternal and fetal infection. Clin. Obstet. Gynaecol., 1974, 1:17-39.

529. Davis, G.L. Cytomegalovirus in the inner ear: case report and electron microscopic study. Ann. Otol. Rhinol. Laryngol., 1969, 78:1179-1188.

530. Davis, L.E., Tweed, G.V., Chin, T.D. and Miller, G.L. Intrauterine diagnosis of cytomegalovirus infection: viral recovery from amniocentesis fluid. Am. J. Obstet. Gynecol., 1971, 109:1217-1219.

531. Davis, L.E., Tweed, G.V. and Stewart, J.A. Cytomegalovirus mononucleosis in a first trimester pregnant female with transmission to the fetus. Pediatrics, 1971, 48:200-206.

532. Davis, L.E., Tweed, G.V., Stewart, J.A., Bernstein, M.T., Miller, G.L., Gravelle, C.R. and Chin, T.D. Cytomegalovirus mononucleosis in a first trimester pregnant female with transmission to the fetus. Pediatrics, 1971, 48:200-206.

533. Daw, E. and Chandler, G. Pregnancy following poliomyelitis. Postgrad. Med. J., 1976, 52:492-496.

534. Dawson, J.B. Antenatal rubella. N. Engl. J. Med., 294:54-55.

535. DCruz, I.A., Balani, S.G. and Iyer, L.S. Infectious hepatitis and pregnancy. Obstet. Gynecol., 1968, 31:449-455.

536. Dehner, L.P. and Askin, F.B. Cytomegalovirus endometritis: report of a case associated with spontaneous abortion. Obstet. Gynecol., 1975, 45:211-214.

537. Deibel, R. Virus diseases of the embryo. Ger. Med. Mon., 1966, 11:416-420.

538. Deinhardt, F. Epidemiology and mode of transmission of viral hepatitis A and B. Am. J. Clin. Pathol., 1976, 65:890-897.

539. Delage, G., Montplaisir, S., Remy-Prince, S. and Pierri, E. Prevalence of hepatitis B virus infection in pregnant women in the Montreal area. Can. Med. Assoc. J., 1986, 134:897-901.

540. Delahunt, C.S. and Rieser, N. Rubella-induced embryopathies in monkeys. Am. J. Obstet. Gynecol., 1967, 99:580-588.

541. Delaney, J.J. Rubella vaccination in the puerperium. Rocky Mt. Med. J., 1971, 68:46-48.

542. Delaplane, D., Yogev, R., Crussi, F. and Shulman, S.T. Fatal hepatitis B in early infancy: the importance of identifying HBsAg-positive pregnant women and providing immunoprophylaxis to their newborns. Pediatrics, 1983, 72:176-180.

543. Della-Porta, A.J., OHalloran, M.L., Parsonson, I.M., Snowdon, W.A., Murray, M.D., Hartley, W.J. and Haughey, K.J. Akabane disease: isolation of the virus from naturally infected ovine foetuses. Aust. Vet. J., 1977, 53:51-52.

544. Deluise, V.P., Cobo, L.M. and Chandler, D. Persistent corneal edema in the congenital rubella syndrome. Ophthalmology, 1983, 90:835-839.

545. DeMendonca, M.J. Why we should all have an SRH test. Br. Dent. J., 1985, 158:319.

546. DeMeurichy, W. and Pensaert, M. Effect of an enterovirus in gilts inseminated with a semen-virus mixture. Zentralbl-Veterinarmed., 1977, 24:97-103.

547. Demian, S.D., Donnelly, W.H., Jr., and Monif, G.R. Coexistent congenital cytomegalovirus and toxoplasmosis in a stillborn. Am. J. Dis. Child, 1973, 125:420-421.

548. Demian, W., Donnelley, J.F. and Monif, G.R.G. Co-existent congenital cytomegalovirus and toxoplasmosis in a stillborn. Am. J. Pathol., 1973, 71:2a.

549. Deng, P., Cutlip, R.C., Lehmkuhl, H.D. and Brogden, K.A. Ultrastructure and frequency of mastitis caused by ovine progressive pneumonia virus infection in sheep. Vet. Pathol., 1986, 23:184-189.

550. DeNicola, L.K. and Hanshaw, J.B. Congenital and neonatal varicella. J. Pediatr., 1979, 94:175-176.

551. Denoyel, G.A., Gaspar, A., Peyramond, D. and Dumont, M. Prolonged excretion of rubella IgM antibody in two pregnant women. Lancet, 1982, 2:214.

552. Deodhar, K.P., Shenoy, C.K., Trivedi, D.R., Joglekar, S.S., Purandare, V.N. and Bhalerao, R.A. Viral hepatitis during pregnancy. (Autopsy study of 45 cases to evaluate the cause of death.) J. Postgrad. Med., 1971, 17:37-42.

553. Deodhar, N.S. Maternal rubella and the risk of congenital malformation in infants -- a review. Indian J. Public Health, 1968, 12:140-148.

554. Derbyshire, J.B. and Caplan, B.A. The isolation and characterization of a strain of infectious bovine rhinotracheitis virus from stillbirth in swine. Can. J. Comp. Med., 1976, 40:252-256.

555. Derso, A., Boxall, E.H., Tarlow, M.J. and Flewett, T.H. Transmission of HBsAg from mother to infant in four ethnic groups. Br. Med. J., 1978, 1:949-952.

556. DeSilva, L.M., Kampfner, G.L., Lister, C.M. and Tobin, J.O. Identification of pregnancies at risk from cytomegalovirus infection. J. Hyg. (Camb.), 1977, 79:347-354.

557. Desmond, M.M., Wilson, G.S., Melnick, J.L. Singer, D.B., Zion, T.E., Rudolph, A.J., Pineda, R.G., MirHashen, Z. and Blattner, R.J. Congenital rubella encephalitis. J. Pediatr., 1967, 71:311-331.

558. Desmonts, G. and Couvreur, J. Congenital toxoplasmosis. A prospective study of 378 pregnancies. New Engl. J. Med., 1974, 290:1110-1116.

559. Desmyter, J. Serological diagnosis of rubella. Br. Med. J., 1970, 4:119.

560. Desmyter, J., South, M.A. and Rawls, W.E. The IgM antibody response in rubella during pregnancy. J. Med. Microbiol., 1971, 4:107-114.

561. DeVirgiliis, S., Frau, F., Sanna, G., Turco, M.P., Figus, A.L., Cornacchia, G. and Cao, A. Perinatal hepatitis B virus detection by hepatitis B virus-DNA analysis. Arch. Dis. Child, 1985, 60:56-58.

562. DeVore, N.E., Jackson, V.M. and Piening, S.L. TORCH infections. Am. J. Nurs., 1983, 83:1660-1665.

563. Dickie, E.R., London, W.T., Merten, C. and Blumberg, B.S. Hepatitis B virus infection in infants of anti-HBS (+) mothers. Lancet, 1981, 1:1000.

564. Dienstag, J.L. Non-A, non-B hepatitis. Adv. Intern. Med., 1980, 26:187-233.

565. Dienstag, J.L. Non-A, non-B hepatitis. I. Recognition, epidemiology, and clinical features. Gastroenterology, 1983, 85:436-462.

566. Dienstag, J.L. and Alter, H.J. Non-A, non-B hepatitis: evolving epidemiologic and clinical perspective. Semin Liver Dis., 1986, 6:67-81.

567. Dietzman, D.E., Madden, D.L., Sever, J.L., Lander, J.J. and Purcell, R.H. Lack of relationship between Down's syndrome and maternal exposure to Australia antigen. Am. J. Dis. Child, 1972, 124:195-197.

568. Dimmick, J., Mahmood, K. and Altshuler, G. Antenatal infection: adequate protection against hyaline membrane disease. Obstet. Gynecol., 1976, 47:56.

569. Dini, M., Alrenga, D.P. and Freese, U. Perinatal Herpes virus infection: report of a case indicating the paternal role. J. Natl. Med. Assoc., 1980, 72:1193-1196.

570. Diosi, P., Babusceac, L., Neviglovschi, O. and DunStoicu, G. Cytomegalovirus infection associated with pregnancy. Lancet, 1967, 2:1063-1066.

571. Diosi, P. and David, C. Cytomegalic inclusion disease: a historical outline. Clio. Medica., 1968, 3:149-166.

572. Disalvo, A.F. Rubella diagnosis. J. S.C. Med. Assoc., 1970, 66:51-53.

573. Diwan, B.A., Meier, H. and Huebner, R.J. Transplacental effects of 1-ethyl-1-nitrosourea in inbred strains of mice. 3. Association between infections of subinfectious endogenous type-C RNA tumor virus expression and chemically induced tumorigenesis. J. Natl. Cancer Inst., 1973, 51:1965-1970.

574. Dixon, R.J., Hartley, W.J., Hutchins, D.R., Lepherd, E.E., Feilen, C., Jones, R.F., Love, D.N., Sabine, M. and Wells, A.L. Perinatal foal mortality associated with a herpesvirus. Aust. Vet. J., 1977, 53:603.

575. Dixon, R.J., Hartley, W.J., Hutchins, D.R., Lepherd, E.E., Feilen, C., Jones, R.F., Love, D.N., Sabine, M. and Wells, A.L. Perinatal foal mortality associated with a herpesvirus. Aust. Vet. J., 1978, 54:103-105.

576. Dodion-Fransen, J., Dekegel, D. and Thiry, L. Congenital varicella-zoster infection related to maternal disease in early pregnancy. Scand. J. Infect. Dis., 1973, 5:149-153.

577. Dodrill, C., MacFarlane, D. and Boyd, R. Effects of intrauterine rubella infection and its consequent physical symptoms on intellectual abilities. J. Consult. Clin. Psychol., 1974, 42:251-255.

578. Doherty, P.C., Vantsis, J.T. and Hart, R. Louping-ill encephalomyelitis in the sheep. VI. Infection of the 120-day foetus. J. Comp. Pathol., 1972, 82:385-392.

579. Doll, R., Hill, A.B. and Sakula, J. Asian influenza in pregnancy and congenital defects. Br. J. Prev. Soc. Med., 1960, 14:167.

580. Donaldson-Wood, C.R., Joo, H.S. and Johnson, R.H. The effect on reproductive performance of porcine parvovirus infection in a susceptible pig herd. Vet. Rec., 1977, 100:237-239.

581. Done, J.T. and Harding, J.D. The relationship of maternal swine fever infection to cerebellar hypoplasia in piglets. Proc. R. Soc. Med., 1966, 59:1083-1084.

582. Done, J.T., Terlecki, S., Richardson, C., Harkness, J.W., Sands, J.J., Patterson, D.S., Sweasey, D., Shaw, I.G., Winkler, C.E. and Duffell, S.J. Bovine virus diarrhoea-mucosal disease virus: pathogenicity for the fetal calf following maternal infection. Vet. Rec., 1980, 106:473-479.

583. Donovan, J., Adelstein, A.M. and Leighton, P. Sequelae of virus infection in pregnancy. Br. Med. J., 1974, 2:502.

584. Donovan, J.W. Chicken pox and pregnancy. Med. J. Aust., 1977, 1:420-421.

585. Dorffel, W. and Nobel, B. Immunity following congenital measles. Scand. J. Infect. Dis., 1972, 4:161-163.

586. Dorfman, S.F. and Bowers, C.H., Jr. Rubella susceptibility among prenatal and family planning clinic populations. Mt. Sinai J. Med. (N.Y.), 1985, 52:248-252.

587. Douglas, J.M., Farmer, R.G., Vidt, D.G. and Thomas, J.P. Hepatitis and renal failure in pregnancy. Report of a patient. Arch. Intern. Med., 1968, 122:59-62.

588. Doxiadis, S., Pantelakis, S. and Valaes, T. Down's syndrome and infectious hepatitis. Lancet, 1970, 1:897.

589. Driscoll, S.G. Histopathology of gestational rubella. Am. J. Dis. Child, 1969, 118:49-53.

590. Droughet, V. and Levantis, F. Le passage transplacentaire du virus Coxsackie B3 chez la souris gestante et la contamination foetale in utero. Ann. Inst. Pasteur, 1968, 114:249.

591. Droughet, V. and Rouquette, C. Serological screening for inapparent Coxsackie B virus infections during pregnancy in mothers of malformed children and in those of normal children. Bull. INSERM, 1970, 25:29.

592. Drucker, J., Barin, F., Chiron, J.P., Coursaget, P. and Goudeau, A. Hepatitis B virus infection in infants of anti-HBs positive mothers. Lancet, 1981, 2:259.

593. Du-Pan, R.M. Protection of pregnant women against rubella. Lancet, 1969, 2:546.

594. Dube, B., Bhattacharya, S., Sinha, V.N., Pandey, L.K. and Dube, R.K. Blood coagulations studies in pregnant patients with infective hepatitis. Br. J. Obstet. Gynaecol., 1977, 84:592-599.

595. Dudgeon, J.A. Virus infections of the human foetus. Proc. R. Soc. Med., 1966, 59:1084-1087.

596. Dudgeon, J.A. A new look at rubella. Proc. R. Soc. Med., 1967, 60:642-648.

597. Dudgeon, J.A. Maternal rubella and its effect on the foetus. Arch. Dis. Child, 1967, 42:110-125.

598. Dudgeon, J.A. Rubella. Clinical aspects of prenatal and postnatal infection. Public Health, 1967, 81:268-279.

599. Dudgeon, J.A. Rubella. Practitioner, 1967, 616-627.

600. Dudgeon, J.A. Dysmorphogenesis: the role of infection. Proc. Roy. Soc. Med., 1968, 61:1285.

601. Dudgeon, J.A. Virus infections. Proc. R. Soc. Med., 1968, 61:995-1000.

602. Dudgeon, J.A. Congenital rubella. Pathogenesis and immunology. Am. J. Dis. Child, 1969, 118:35-44.

603. Dudgeon, J.A. Immunization against rubella. Nature, 1969, 223:674-676.

604. Dudgeon, J.A. Prenatal virus infections: In: Debre and R. Celers (eds.). The Evaluation and Management of Human Viral Infections. Philadelphia: Saunders, 1970.

605. Dudgeon, J.A. Prevention of congenital rubella defects. The need for long-term surveillance. Arch. Dis. Child, 1970, 45:820.

606. Dudgeon, J.A. Teratogenic effect of rubella virus. Proc. R. Soc. Med., 1970, 63:1254-1257.

607. Dudgeon, J.A. Virological aspects of intra-uterine infection. J. Pathol., 1970, 101:Pxii.

608. Dudgeon, J.A. Future developments in prophylaxis. Ciba. Found. Symp., 1972, 10:179-198.

609. Dudgeon, J.A. Congenital rubella in the United Kingdom of Great Britain. In: Krugman, S., Gershon, A.A., ed. Infections of the fetus and the newborn infant. New York, Liss, 1975, 23-34.

610. Dudgeon, J.A. Congenital rubella. J. Pediat., 1975, 87:1078.

611. Dudgeon, J.A. Intra-uterine infection. Proc. Roy. Soc. Med., 1975, 68:365.

612. Dudgeon, J.A. Viral infections. J. Clin. Pathol., 1976, 10:99-106.

613. Dudgeon, J.A., Marshall, W.C. and Soothill, J.F. Immunological responses to early and late intrauterine virus infections. J. Pediatr., 1969, 75:1149-1166.

614. Duehr, P.A. Herpes zoster as a cause of congenital cataract. Am. J. Ophthalmol., 1955, 39:157.

615. Duffell, S.J., Sharp, M.W., Winkler, C.E., Terlecki, S., Richardson, C., Done, J.T., Roeder, P.L. and Hebert, C.N. Bovine virus diarrhoea-mucosal disease virus-induced fetopathy in cattle: Efficacy of prophylactic maternal pre-exposure.

616. Dukes, P.C. Rubella immunity. N.Z. Med. J., 1983, 96:266.

617. Duncan, J.R. and Kenney, A. Rubella embryopathy. Br. Med. J., 1978, 2:504.

618. Dundas, G.H.G. Ophthalmia neonatorum before birth. Lancet, 1921, 1:122.

619. Dunlevy, M. Rubella contact in pregnancy. J. Ir. Med. Assoc., 1973, 66:131-134.

620. Dunlevy, M. A study into the epidemiology of rubella defects. Ir. Med. J., 1975, 68:478-481.

621. Dunn, A.E., Peters, R.L. and Schweitzer, I.L. An ultrastructural study of liver explants from infants with vertically transmitted hepatitis. Exp. Mol. Pathol., 1973, 19:113-126.

622. Dunne, H.W., Ajinkya, S.M., Bubash, G.R. and Griel, L.C., Jr. Parainfluenza-3 and bovine enteroviruses as possible important causative factors in bovine abortion. Am. J. Vet. Res., 1973, 34:1121-1126.

623. Dunne, H.W. and Clark, C.D. Embryonic death, fetal mummification, stillbirth, and neonatal death in pigs of gilts vaccinated with attenuated live-virus hog cholera vaccine. Am. J. Vet. Res., 1968, 29:787-796.

624. Dunne, H.W., Gobble, J.L., Hokanson, J.F., Kradel, D.C. and Bubash, G.R. Porcine reproductive failure associated with a newly identified "SMEDI" group of picorna viruses. Am. J. Vet. Res., 1965, 26:1284-1297.

625. Dunne, H.W., Wang, J.T., Clark, C.D., Hokanson, J.F., Morimoto, T. and Bubash, G.R. The effects of in utero viral infection on embryonic, fetal, and neonatal survival: a comparison of SMEDI (porcine picorna) viruses with hog cholera vaccinal virus. Can. J. Comp. Med., 1969, 33:244-252.

626. Dunne, H.W., Wang, J.T. and Huang, C.M. Early in utero infection of porcine embryos and fetuses with SMEDI (entero-) viruses: mortality, antibody development, and viral persistence. Am. J. Vet. Res., 1974, 35:1479-1481.

627. Duong, C.M. and Munns, R.E. Varicella pneumonia during pregnancy. J. Fam. Pract., 1979, 8:277-280.

628. Dupont, C. Herpes gestationis with hydatidiform mole. Trans. St. Johns Hosp. Dermatol. Soc., 1974, 60:103.

629. Dupuy, J.M. Viral hepatitis and autoimmune liver diseases. Boll. Ist. Sieroter. Milan, 1974, 53:221-225.

630. Dupuy, J.M., Frommel, D. and Alagille, D. Severe viral hepatitis type B in infancy. Lancet, 1975, 1:191-194.

631. Duran-Reynals, M.L. Maternal transmission of resistance to virus-induced leukemia. Prog. Med. Virol., 1985, 47-57.

632. Durham, P.J. Infectious bovine rhinotracheitis virus and its role in bovine abortion. N.Z. Vet. J., 1974, 22:175-180.

633. Durham, P.J., Forbes-Faulkner, J.C. and Poole, W.S. Infectious bovine rhinotracheitis virus: experimental attempts at inducing bovine abortion with a New Zealand isolate. N.Z. Vet. J., 1975, 23:93-94.

634. Durodola, J.I. Administration of cyclophosphamide during late pregnancy and early lactation: a case report. J. Natl. Med. Assoc., 1979, 71:165-166.

635. Dutta, D. Perinatal nonbacterial infections. Clin. Lab. Med., 1981, 1:377-397.

636. Dutta, S.K. and Campbell, D.L. Cell mediated immunity in equine herpesvirus type 1 infection I. In vitro lymphocyte blastogenesis and serum neutralization antibody in normal parturient and aborting mares. Can. J. Comp. Med., 1977, 41:404-408.

637. Dyer, I. Measles complicating pregnancy. Report of 24 cases with three instances of congenital measles. South Med. J., 1940, 33:601.

638. Dykes, M.H. an Meier, P. Ascorbic acid and the common cold. Evaluation of its efficacy and toxicity. J.A.M.A., 1975, 231:1073-1079.

# E


639. Eachempati, U. and Woods, R.E. Cytomegalic virus disease in pregnancy. Obstet. Gynecol., 1976, 47:615-618.

640. Easterly, J.R. and Oppenheimer, E.H. Intra-uterine rubella infection. Perspect. Pediatr. Res., 1973, 1:313.

641. Ebbesen, P. A subcellular agent inducing plasma cell leukemias in mice. The relation of this agent to amyloidosis and chronic murine pneumonia. Acta. Pathol. Microbiol. Scand., 1968, 197:9.

642. Echevarria, C., Fernandez, M.V., Echevarria, J.M. and Najera, R. Pregnant women in contact with rubella. J. Infect., 1985, 11:80-82.

643. Eckroade, R.J., ZuRhein, G.M. and Hanson, R.P. Transmissible mink encephalopathy in carnivores: clinical, light and electron microscopic studies in raccons, skunks and ferrets. J. Wildl. Dis., 1973, 9:229-240.

644. Eden, A.N. Perinatal morbidity caused by Vibrio fetus. J. Pediat., 1966, 68:297.

645. Edington, N., Watt, R.G. and Plowright, W. Experimental transplacental transmission of porcine cytomegalovirus. J. Hyg. (Camb.), 1977, 78:243-251.

646. Edmond, E. and Zealley, H. The impact of a rubella prevention policy on the outcome of rubella in pregnancy. Br. J. Obstet. Gynaecol., 1986, 93:563-567.

647. Edwards, K.R., Emmerson, M.A., Luff, P.R., Wells, D.E., Muskett, J.C., Wrathall, A.E., Richardson, C., Parker, B.N. and Thornton, D.H. Efficacy of porcine parvovirus vaccines. Vet. Rec., 1986, 119:203-206.

648. Edwards, M.J. Influenza, hyperthermia, and congenital malformation. Lancet, 1972, 1:320-321.

649. Edwards, M.S. Venereal herpes: a nursing overview. J.O.G.N. Nurs., 1978, 7:7-15.

650. Eichenwald, H.F., McCraken, G.H., Jr., and Kindberg, S.J. Virus infections of the newborn. Prog. Med. Virol., 1967, 9:35-104.

651. Eilard, T. and Strannegard, O. Rubella reinfection in pregnancy followed by transmission to the fetus. J. Infect. Dis., 1974, 129:594-596.

652. Elek, S.D. and Stern, H. Development of a vaccine against mental retardation caused by cytomegalovirus infection. Lancet, 1974, 1:1-4.

653. Elek, S.D. and Stern, H. Development of a vaccine against mental retardation caused by cytomegalovirus infection in utero. Lancet, 1974, 2:1-5.

654. Eliezer, S., Ester, F., Ehud, W. and Henryk, Z. Fetal splenomegaly, ultrasound diagnosis of cytomegalovirus infection: a case report. J.C.U., 1984, 12:520-521.

655. Elizan, T.S., Ajero-Froehlich, L., Fabiyi, A., Ley, A. and Sever, J.L. Viral infection in pregnancy and congenital CNS malformations in man. Arch. Neurol., 1969, 20:115-119.

656. Elliott, G.B., McAllister, J.E. and Alberta, C. Fetal poliomyelitis. Amer. J. Obst. Gyn., 1956, 72:896.

657. Elliott, L.P. and Amplatz, K. Angiocardiographic observations in the post-rubella syndrome. Preliminary report. Am. J. Roentgenol. Radium. Ther. Nucl. Med., 1966, 97:164-166.

658. Ellis, T.M. Jaundice in a Siamese cat with in utero feline calicivirus infection. Aust. Vet. J., 1981, 57:383-385.

659. Ellis, W.A., Bryson, D.G. and McFerran, J.B. Abortion associated with mixed Leptospira/equid herpesvirus 1 infection. Vet. Rec., 1976, 98:218-219.

660. Ellman, R. Rubella immunisation. Br. Med. J., 1979, 2:1440-1441.

661. Embil, J.A., Haldane, E.V., MacKenzie, R.A. and Rooyen, C.E., Van. Prevalence of cytomegalovirus infection in a normal urban population in Nova Scotia. Can. Med. Assoc. J., 1969, 101:78-81.

662. Embil, J.A., Haldane, E.V. and OShaughnessy, M.V. A five-year study of cytomegaloviruria in an adult. Can. J. Public Health, 1972, 63:249-256.

663. Embil, J.A., Ozere, R.L. and Haldane, E.V. Congenital cytomegalovirus infection in two siblings from consecutive pregnancies. J. Pediatr., 1970, 77:417-421.

664. Emerson, J.L. and Delez, A.L. Prenatal hog cholera infection: a potential source of hog cholera. J. Am. Vet. Med. Assoc., 1965, 147:1346-1349.

665. Enders, G. Serodiagnosis of Varicella-Zoster virus infection in pregnancy and standardization of the ELISA IgG and IgM antibody tests. Dev. Biol. Stand., 1982, 52:221-236.

666. Enders, G. Varicella-zoster virus infection in pregnancy. Prog. Med. Virol., 1984, 29:166-196.

667. Enders, G. Management of varicella-zoster contact and infection in pregnancy using a standardized varicella-zoster ELISA test. Postgrad. Med. J., 1985, 61:23-30

668. Enders, G. Rubella antibody titers in vaccinated and nonvaccinated women and results of vaccination during pregnancy. Rev. Infect. Dis., 1985, 1:103-107.

669. Enders, G., Calm, A. and Schaub, J. Rubella embryopathy after previous maternal rubella vaccination. Infection, 1984, 12:96-98.

670. Enright, F.M. and Osburn, B.I. Ontogeny of host responses in ovine fetuses infected with bluetongue virus. Am. J. Vet. Res., 1980, 41:224-229.

671. Epstein, L.G., Sharer, L.R., Joshi, V.V., Fojas, M.M., Koenigsberger, M.R. and Oleske, J.M. Progressive encephalopathy in children with acquired immune deficiency syndrome. Annals of Neurology, 1985, 17:488-496.

672. Erdelyi, R. The influence of toxoplasmosis on the incidence of congenital facial malformations, preliminary report. Plast. Reconstr. Surg., 1957, 20:306-310.

673. Erkman, J. and Blythe, J.G. Azathioprine therapy complicated by pregnancy. Obstet. Gynecol., 1972, 40:708-710.

674. Ernyei, S. Contributions to the question of the teratogenesis of mongolism. Eye, Ear, Nose, Throat, Mon., 1968, 47:271-273.

675. Essex, M. Horizontally and vertically transmitted oncornaviruses of cats. Adv. Cancer Res., 1975, 21:175-248.

676. Esterly, J.R. and Oppenheimer, E.H. Vascular lesions in infants with congenital rubella. Circulation, 1967, 36:544-554.

677. Esterly, J.R. and Oppenheimer, E.H. Pathological lesions due to congenital rubella. Arch. Pathol., 1969, 87:380-388.

678. Esterly, J.R. and Oppenheimer, E.H. Intrauterine rubella infection. Perspect. Pediatr. Pathol., 1973, 1:313-338.

679. Euler, A.R. Mother-infant transmission of hepatitis B antigen. Am. J. Dis. Child, 1976, 130:566.

680. Evans, E.B., Pollock, T.M., Cradock-Watson, J.E. and Ridehalgh, M.K. Human anti-chickenpox immunoglobulin in the prevention of chickenpox. Lancet, 1980, 1:354-356.

681. Evans, J.M. and Forrest, D.T. Prevention of canine distemper on infected premises. J. Am. Vet. Med. Assoc., 1970, 156:1758-1762.

682. Evans, T.J., McCollum, J.P. and Valdimarsson, H. Congenital cytomegalovirus infection after maternal renal transplantation. Lancet, 1975, 1:1359-1360.

683. Evermann, J.F. and Trueblood, M.S. Characteristics of a paramyxovirus isolated from an aborted bovine fetus. Cornell Vet., 1973, 63:17-28.

684. Evrard, J.R. Rubella, the Trojan horse of obstetrics. Obstet. Gynecol., 1981, 58:505-507.

685. Ewing, C.I. and Davidson, D.C. Fatal hepatitis B in infant born to a HBsAg carrier with HNeAb. Arch. Dis. Child, 1985, 60:265-267.

686. Eyal, A., Friedman, M., Peretz, B.A. and Paldi, E. Pregnancy complicated by herpes zoster. A report of two cases and literature review. J. Reprod. Med., 1983, 28:600-603.

# F


687. Fair, J.R. Congenital toxoplasmosis. IV. Case finding using the skin test and ophthalmoscope in state schools for mentally retarded children. Amer. J. Ophthalmol., 1959, 48:813-819.

688. Fair, J.R. Congenital toxoplasmosis. V. Ocular aspects of the disease. J. Med. Assoc. Ga., 1959, 48:604-607.

689. Fairchild, B. But don't quote me: reflections by parents on deafness and rubella. Am. Ann. Deaf., 1980, 125:1007-1011.

690. Faiz, R.G., Zweig, S.E., Kummer, J.F., Moore, D. and Lang, D.J. Cytomegalovirus-specific cell-mediated immunity during pregnancy in lower socioeconomic class adolescents. J. Infect. Dis., 1983, 148:621-629.

691. Farber, P.A. and Glasgow, L.A. Viral myocarditis during pregnancy: encephalomyocarditis virus infection in mice. Am. Heart J., 1970, 80:96-102.

692. Farquhar, H.G. Congenital toxoplasmosis. Report of two cases in twins. Lancet, 1950, 259:562-564.

693. Farquhar, J.D. Follow-up on rubella vaccinations and experience with subclinical reinfection. J. Pediatr., 1972, 81:460-465.

694. Farwell, J.R., Dohrmann, G.J., Marrett, L.D. and Meigs, J.W. Effect of SV40 virus-contaminated polio vaccine on the incidence and type of CNS neoplasms in children: a population-based study. Trans. Am. Neurol. Assoc., 1979, 104:261-264.

695. Fawaz, K.A., Grady, G.F., Kaplan, M.M. and Gellis, S.S. Repetitive maternal-fetal transmission of fatal hepatitis B. N. Engl. J. Med., 1975, 293:1357-1359.

696. Fedrick, J. and Alberman, E.D. Reported influenza in pregnancy and subsequent cancer in the child. Br. Med. J., 1972, 2:485-488.

697. Feeman, W.E., Jr. Clinical rubella after reinfection. N. Engl. J. Med., 1973, 288:525.

698. Feldman, H.A. and Miller, L.T. Congenital human toxoplasmosis. Ann. N.Y. Acad. Sci., 1956, 64:180-184.

699. Feldman, R.A. Cytomegalovirus infection during pregnancy. A prospective study and report of six cases. Am. J. Dis. Child, 1969, 117:517-521.

700. Feldman, R.B., Pinsky, L., Mendelson, J. and Lajoie, R. Can language disorder not due to peripheral deafness be an isolated expression of prenatal rubella? Pediatrics, 1973, 52:296-299.

701. Feldman, S. Varicella zoster infections of the fetus neonata and immunocompromised child. In: Aronoff, S.C. (ed.) Advances in Pediatric Infectious Diseases, Year Book Medical Publishers, Inc., Chicago, Ill., USA; London, England, 1986, 99:116.

702. Felman, Y.M. and Nikitas, J.A. Cytomegalovirus infection. Cutis., 1981, 27:562-568.

703. Felsburg, P.J., Heberling, R.I., Brack, M. and Kalter, S.S. Experimental genital herpes infection in the marmoset. J. Med. Prim., 1973, 2:50.

704. Felsburg, P.J., Heberling, R.L. and Kalter, S.S. Experimental genital infection of cebus monkeys with oral and genital isolates of herpesvirus hominis type 1 and type 2. Arch. ges. Virusforsch., 1972, 39:223.

705. Ferenczy, A. Treating genital condyloma during pregnancy with the carbon dioxide laser. Am. J. Obstet. Gynecol., 1984, 148:9-12.

706. Ferguson, J.A., Reeves, W.C. and Hardy, J.L. Studies on immunity to alphaviruses in foals. Am. J. Vet. Res., 1979, 40:5-10.

707. Ferm, V.H. and Kilham, L. Mumps virus infection in the pregnant hamster. J. Embryol. Exp. Morphol., 1963, 11:659.

708. Ferm, V.H. and Kilham, L. Congenital anomalies induced in hamster embryos with H-1 virus. Science, 1964, 145:510-511.

709. Fernelius, A.L., Amtower, W.C., Malmquist, W.A., Lambert, G. and Matthews, P.J. Bovine viral diarrhea virus in swine: neutralizing antibody in naturally and experimentally infected swine. Can. J. Comp. Med., 1973, 37:96-102.

710. Ferrer, J.F., Piper, C.E., Abt, D.A., Marshak, R.R. and Bhatt, D.M. Natural mode of transmission of the bovine C type leukemia virus (BLV). Bibl. Haematol., 1975, 43:235-237.

711. Figueroa, F., Ohlbaum, A. and Contreras, G. Neutralizing antibody response in bovine serum and nasal and salivary secretions after immunization with live or inactivated foot-and-mouth disease virus. Infect. Immun., 1973, 8:296-298.

712. Filloux, F., Kelsey, D.K., Bose, C.L., Veasy, L.G. and Gooch, W.M. III. Hydrops fetalis with supraventricular tachycardia and cytomegalovirus infection. Clin. Pediatr. (Phila.), 1985, 24:534-536.

713. Fine, P.E., Adelstein, A.M., Snowman, J., Clarkson, J.A. and Evans, S.M. Long term effects of exposure to viral infections in utero. Br. Med. J., 1985, 290:509-511.

714. Finkel, A., Dent, P.B., Emrich, W.H., Gent, M. and Rahim, M.A. Intra-uterine infection and cord immunoglobulin M. II. Clinical analysis of infants with elevated cord serum immunoglobulin M. Can. Med. Assoc. J., 1974, 110:38-42.

715. Finkle, H.I. and Goldman, R.L. Burkitt's lymphoma -- gynecologic considerations. Obstet. Gynecol., 1974, 43:281-284.

716. Finelli, P.F. Herpes simplex virus and the human nervous system: current concepts and review. Milit. Med., 1975, 140:765-771.

717. Firkin, F.C., Nicholls, K. and Whelan, G. Transient myeloid and erythroid aplasia associated with infectious hepatitis. Br. Med. J., 1978, 2:1534.

718. Fisher, S. Rubella immunization: the problem of non-converters. Med. J. Aust., 1971, 2:1094.

719. Fitzgerald, J.F., Angelides, A. and Wyllie, R. The hepatitis spectrum. Curr. Probl. Pediatr., 1981, 11:1-51.

720. Fitzgerald, M.D., Sitton, A.B. and McConnell, F. Audiometric, developmental, and learning characteristics of a group of rubella deaf children. J. Speech Hear Disord., 1970, 35:218-228.

721. Fiumara, N.J. and Lessell, S. Manifestations of late congenital syphilis. An analysis of 271 patients. Arch. Dermatol., 1970, 102:78.

722. Flanagan, M., Wilson, A.J., Trueman, K.F. and Shepherd, M.A. Bluetongue virus serotype 20 infection in pregnant Merino sheep. Aust. Vet. J., 1982, 59:18-20.

723. Flanders, R.W., Person, D.A., Orren, J.M., Adams, J.M., Jr. and Rawls, W.E. Is herpes simplex infection of the newborn preventable? The possible role of cesarean section in minimizing risk. Clin. Pediatr. (Phila.), 1972, 11:293-294.

724. Fleet, W.F., Jr., Benz, E.W., Jr., Karzon, D.T., Lefkowitz, L.B. and Herrmann, K.L. Fetal consequences of maternal rubella immunization. J.A.M.A., 1974, 227:621-627.

725. Fleet, W.F., Vaughn, W., Lefkowitz, L.B., Jr., Schaffner, W. and Federspiel, C.F. Gestational exposure to rubella vacines: a population surveillance study. Am. J. Epidemiol., 1975, 101:220-230.

726. Fleisher, G. and Bolognese, R. Persistent Epstein-Barr virus infection and pregnancy. J. Infect. Dis., 1983, 147:982-986.

727. Fleisher, G. and Bologonese, R. Epstein-Barr virus infections in pregnancy: a prospective study. J. Pediatr., 1984, 104:374-379.

728. Fleisher, G. and Bologonese, R. Infectious mononucleosis during gestation: report of three women and their infants studied prospectively. Pediatr. Infect. Dis., 1984, 3:308-311.

729. Florman, A.L., Gershon, A.A., Blackett, P.R. and Nahmias, A.J. Intrauterine infection with herpes simplex virus. Resultant congenital malformations. J.A.M.A., 1973, 225:129-132.

730. Fogel, A., Gerichter, C.B. and Goldblum, N. Prevention of the congenital rubella syndrome: the Israeli experience. Prog. Clin. Biol. Res., 1985, 327-332.

731. Fogel, A., Gerichter, C.B., Handsher, R., Bernholtz-Barnea, B. The impact of rubella epidemics on pregnant women, proof for the necessity of pre-marital screening and vaccination. Dev. Biol. Stand., 1979, 43:349-353.

732. Fogel, A., Gerichter, C.B., Rannon, L., Bernholtz, B. and Handsher, R. Serologic studies in 11,460 pregnant women during the 1972 rubella epidemic in Israel. Am. J. Epidemiol., 1976, 103:51-59.

733. Fogel, A., Handsher, R. and Barnea, B. Subclinical rubella in pregnancy -- occurrence and outcome. Isr. J. Med. Sci., 1985, 21:133-138.

734. Forbes, P.B. The significance of AIDS in obstetric practice. Br. J. Hosp. Med., 1986, 36:342-346.

735. Forman, A.J., Lenghaus, C., Hogg, G.G. and Hale, C.J. Association of a parvovirus with an outbreak of foetal death and mummification in pigs. Aust. Vet. J., 1977, 53:326-329.

736. Forrest, J.H., Honeyman, M.C. and Murphy, A.M. Gamma globulin and maternal rubella. Med. J. Aust., 1973, 1:745-746.

737. Forrest, J.M. and Menser, M.A. Congenital rubella in school children and adolescents. Arch. Dis. Child, 1970, 45:63.

738. Forrest, J.M. and Menser, M.A. Storage of rubella serum. Med. J. Aust., 1971, 2:779.

739. Forrest, J.M. and Menser, M.A. Gammaglobulin and congenital rubella. Br. Med. J., 1974, 2:439.

740. Forrest, J.H. and Menser, M.A. Recent implications of intra-uterine and postnatal rubella. Aust. Paediat. J., 1975, 11:65-75.

741. Forrest, J.M. and Menser, M.A. Recent implications of intrauterine and postnatal rubella. Aust. Paediatr. J., 1975, 11:65-75.

742. Forrest, J.H. and Menser, M.A. Failure of rubella vaccination to prevent congenital rubella. Med. J. Aust., 1977, 1:77.

743. Forrest, J.M., Menser, M.A. and Burgess, J.A. High frequency of diabetes mellitus in young adults with congenital rubella. Lancet, 1971, 2:332-334.

744. Forrest, J.M., Menser, M.A. and Harley, J.D. Diabetes mellitus and congenital rubella. Pediatrics, 1969, 44:445-446.

745. Forrest, R.D., Bostrom, H. and Dahlberg, P.A. IgA deficiency during penicillamine treatment. Br. Med. J., 1977, 1:777.

746. Forrester, R.M., Lees, V.T. and Watson, G.H. Rubella syndrome: escape of a twin. Brit. Med. J., 1966, 1:1403.

747. Forsgren, M., Carlstrom, G. and Stranger, K. Congenital rubella after maternal reinfection. Scand. J. Infect. Dis., 1979, 11:81-83.

748. Forsgren, M. and Soren, L. Subclinical rubella reinfection in vaccinated women with rubella-specific IgM response during pregnancy and transmission of virus to the fetus. Scand. J. Infect. Dis., 1985, 17:337-341.

749. Forster, R.K., Dawson, C.R. and Schachter, J. Late follow-up of patients with neonatal inclusion conjunctivitis. Am. J. Ophthalmol., 1970, 69:467-472.

750. Fortuin, N.J., Morrow, A.G. and Roberts, W.C. Late vascular manifestations of the rubella syndrome. A roentgenographic-pathologic study. Am. J. Med., 1971, 51:134-140.

751. Foschini, M., DeToni, A., Padula, D., Pizzocolo, G., Arrighini, A. and Zunin, C. Prevention of perinatally transmitted hepatitis B virus infection by hepatitis B immunoglobulin immunoprophylaxis: an account of 201 newborn babies of hepatitis Bs antigen carrier mothers. J. Pediatr. Gastroenterol. Nutr., 1985, 4:523-527.

752. Fox, J.L. Injected virus probes fetal development. Science, 1984, 223:1377.

753. Fox, J.P. Fetal perspective of rubella vaccine efficacy and safety. J.A.M.A., 1981, 246:1444-1445.

754. Fox, M.J., Krumpiegel, E.R. and Teresi, J.L. Maternal measles, mumps, and chickenpox as a cause of congenital anomalies. Lancet, 1948, 1:746.

755. Frampton, J. Rubella immunisation. Br. Med. J., 1978, 2:698.

756. Franciosi, R.A. Viruses and congenital malformations. N. Engl. J. Med., 1972, 287:1099-1100.

757. Francis, B.H., Hatherley, L.I., Walstab, J.E. and Taft, L.I. Rubella screening and vaccination programme at a Melbourne maternity hospital. A five-year review. Med. J. Aust., 1982, 1:502-504.

758. Francis, D.P. Worldwide control of hepatitis B virus: an approaching reality? Pediatrics, 1985, 76:851-852.

759. Francis, M.J. and Black, L. Effect of the sow vaccination regimen on the decay rate of maternally derived foot-and-mouth disease antibodies in piglets. Res. Vet. Sci., 1984, 37:72-76.

760. Franco, S.A., Riley, H.D., Jr. and Chitwood, L.A. The congenital rubella syndrome. South Med. J., 1970, 63:825-830.

761. Franzen, C. and Frosner, G. Placental transfer of hepatitis A antibody. N. Engl. J. Med., 1981, 304:427.

762. Freedman, P.S. ECHOvirus 11 infection and intrauterine death. Lancet, 1979, 1:96-97.

763. Freeman, D.W. and Barno, A. Deaths from Asian influenza associated with pregnancy. Am. J. Obstet. Gynecol., 1959, 78:1172.

764. Freestone, D.S. Programmes for the prevention of rubella during pregnancy by active immunization. Br. J. Prev. Soc. Med., 1974, 28:258-264.

765. French, E.L., Hore, D.E., Snowdon, W.A., Parsonson, I.M. and Uren, J. Infection of pregnant ewes with mucosal disease virus of ovine origin. Aust. Vet. J., 1974, 50:45-54.

766. French, M.L., Thompson, J.F. and White, A. Cytomegalovirus viremia with transmission from mother to fetus. Ann. Intern. Med., 1977, 86:748-749.

767. Frenkel, L.D., Crawford, M.A. and Bellanti, J.A. Virological and serological studies in pregnancy, in women with suspect genital herpes simplex virus or perinatal cytomegalovirus infections, and in women with carcinoma of the cervix. I.A.R.C. Sci. Publ., 1978, 911-915.

768. Freund, P., Rook, G.D. and Gurian, S. Herpes zoster in the newborn. Amer. J. Dis. Child, 1942, 64:895.

769. Frey, H.M., Bialkin, G. and Gerson, A.A. Congenital varicella: case report of a serologically proved long-term survivor. Pediatrics, 1977, 59:110-112.

770. Frey, H.R., Liess, B., Richter-Reichhelm, H.B., Von-Benten, K. and Trautwein, G. Experimental transplacental transmission of hog cholera virus in pigs. I. Virological and serological studies. Zentralbl. Veterinarmed., 1980, 27:154-164.

771. Friart, G.M. Vaccination antivariolique in utero. Arch. Belges Derm., 1963, 19:191.

772. Friedman, A.D., Harrison, H.R. and Alexander, E.R. Coinfection with cytomegalovirus and Chlamydia trachomatis: seroepidemiologic data. J. Infect. Dis., 1983, 148:332.

773. Friedmann, I. and Wright, M.I. Histopathological changes in the foetal and infantile inner ear caused by maternal rubella. Br. Med. J., 1966, 2:20-23.

774. Friedrich, E.G., Jr. Reversible vulvar atypia. A case report. Obstet. Gynecol., 1972, 39:173-181.

775. Friendly, D.S. Gonococcal conjunctivitis in the newborn. Clin. Proc. Child Hosp. D.C., 1969, 25:1-9.

776. Frost, J.O. and Miller, M. Vestibular function and hearing in children with prenatal rubella. N.Y. State J. Med., 1971, 71:971-973.

777. Fuccillo, D.A. Congenital varicella. Teratology, 1977, 15:329-330.

778. Fucillo, D.A. and Sever, J.L. Viral teratology. Bacteriol. Rev., 1973, 37:19.

779. Fujisaki, Y., Miura, Y., Sugimori, T., Morimoto, T. and Ino, T. Experimental studies on vertical infection of mice with Japanese encephalitis virus. I. Effect of mouse strain on placental and fetal infection. Natl. Inst. Anim. Health Q. (Tokyo), 1976, 16:97-106.

780. Fujisaki, Y., Miura, Y., Sugimori, T., Morimoto, T. and Ino, T. Experimental studies on vertical infection of mice with Japanese encephalitis virus. II. Effect of inoculation route on placental and fetal infection. Natl. Inst. Anim. Health Q. (Tokyo), 1977, 17:39-44.

781. Fujisaki, Y., Miura, Y., Sugimori, T., Murakami, Y., Ino, T. and Miura, K. Experimental studies on vertical infection of mice with Japanese encephalitis virus. III. Effect of gestation days at the time of inoculation on placental and fetal infection. Natl. Inst. Anim. Health Q. (Tokyo), 1982, 22:95-101.

782. Fujisaki, Y., Miura, Y., Sugimori, T., Murakami, Y. and Miura, K. Experimental studies on vertical infection of mice with Japanese encephalitis virus. IV. Effect of virus strain on placental and fetal infection. Natl. Inst. Anim. Health Q. (Tokyo), 1983, 23:21-26.

783. Fujisaki, Y., Sugimori, T., Morimoto, T. and Miura, Y. Development of an attenuated strain for Japanese encephalitis live virus vaccine for porcine use. Natl. Inst. Anim. Health Q. (Tokyo), 1975, 15:15-23.

784. Fujita, K., Yamada, T., Mayumi, M., Yamashita, Y. and Imai, M. Materno-fetal transport of Australia/S.H. antigen and antibody. Lancet, 1972, 2:378-379.

785. Fukumi, H. and Takeuchi, Y. Vaccination against parainfluenza 1 virus (typus muris) infection in order to eradicate this virus in colonies of laboratory animals. Dev. Biol. Stand., 1975, 28:477-481.

786. Fulginiti, V.A. and Alexander, E.R. Rubella in pregnancy: a guide to rational counseling. J.A.M.A., 1981, 245:1666.

787. Fuller, D.A. Field observations of the efficacy of TGE vaccine. Vet. Med. Small Anim. Clin., 1971, 66:1206.

788. Fung, J.C. and Tilton, R.C. TORCH serologies and specific lgM antibody determination in acquired and congenital infections. Ann. Clin. Lab. Sci., 1985, 15:204-211.

789. Furhoff, A.K. and Hellstrom, K. Jaundice in pregnancy. A follow-up study of the series of women originally reported by L. Thorling. I. The pregnancies. Acta. Med. Scand., 1973, 193:259-266.

790. Furukawa, T., Miyata, T., Kondo, K., Kuno, K., Isomura, S. and Takekoshi, T. Clinical trials of RA 27-3 (Wistar) rubella vaccine in Japan. Am. J. Dis. Child, 1969, 118:262-263.

# G

791. Gagnon, A.N. and Dulac, G.C. Porcine parvovirus infection in Ontario: incidence and diagnosis in herds with reproductive failures. Can. Vet. J., 1979, 20:338.

792. Gagnon, R.A. Transplacental inoculation of fatal herpes simplex in the newborn. Report of 2 cases. Obstet. Gynecol., 1968, 31:682-684.

793. Gaines, K.J., Jabbour, J.T., Whitaker, J.N. and Sever, J. Subacute sclerosing panencephalitis during pregnancy. A surviving normal infant. Arch. Neurol., 1979, 36:314-316.

794. Galbert, M.W. and Gardner, A.E. The anesthetic and obstetric management of the parturient with acute hepatitis. Anesth. Analg., 1972, 51:549-552.

795. Galbraith, R.M., Eddleston, A.L., Williams, R. and Zuckerman, A.J. Fulminant hepatic failure in leukaemia and choriocarcinoma related to withdrawal of cytotoxic drug therapy. Lancet, 1975, 2:528-530.

796. Gallagher, M.R., Welliver, R., Yamanaka, T., Eisenberg, B., Sun, M. and Ogra, P.L. Cell-mediated immune responsiveness to measles. Its occurrence as a result of naturally acquired or vaccine-induced infection and in infants of immune mothers. Am. J. Dis. Child, 1981, 135:48-51.

797. Galloway, P.G. and Roessmann, U. Diffuse dysplasia of cerebral hemispheres in a fetus possible viral cause. Arch. Pathol. Lab. Med., 1987, 111:143-145.

798. Garcia, A.G. Maternal herpes-simplex infection causing abortion. Histopathologic study of the placenta. Hospital, 1970, 78:1267-1274.

799. Garcia, A.G., Marques, R.L., Lobato, Y.Y., Fonseca, M.E. and Wigg, M.D. Placental pathology in congenital rubella. Placenta, 1985, 6:281-295.

800. Garcia, A.G., Pereira, J.M., Vidigal, N., Lobato, Y.Y., Pegado, C.S. and Branco, J.P. Intrauterine infection with mumps virus. Obstet. Gynecol., 1980, 56:756-759.

801. Garcia, A.G.P. Congenital toxoplasmosis in two successive sibs. Arch. Dis. Child, 1968, 43:705-709.

802. Garcia-Tamayo, J., Esparza, J. and Martinez, A.J. Placental and fetal alterations due to Venezuelan equine encephalitis virus in rats. Infect. Immun., 1981, 32:813-821.

803. Gardiner, A.C. The distribution and significance of border disease viral antigen in infected lambs and foetuses. J. Comp. Pathol., 1980, 90:513-518.

804. Gardner, M.B., Arnstein, P., Johnson, E., Rongey, R.W., Charman, H.P. and Huebner, R.J. Feline sarcoma virus tumor induction in cats and dogs. J. Am. Vet. Med. Assoc., 1971, 158:1046.

805. Gardner, M.B., Rasheed, S., Rongey, R.W., Charman, H.P., Alena, B., Gilden, R.V. and Huebner, R.J. Natural expression of feline type-C virus genomes, prevalence of detectable felv and RD- 114 GC antigen, type-C particles and infectious virus in postnatal and fetal cats. Int. J. Cancer, 1974, 14:97-105.

806. Gardner, P.S. Virus infections and respiratory disease of childhood. Arch. Dis. Child, 1968, 43:629-645.

807. Garry, R. and Jones, R. Relationship between cervical condylomata, pregnancy and subclinical papillomavirus infection. J. Reprod. Med., 1985, 30:393-399.

808. Gartner, L., Kunkel, M., Oberender, H. and Klausch, B. Comparison of different serological methods in diagnosing primary and recurrent cytomegalovirus infections in pregnancy. Acta. Virol. (Praha.), 1979, 23:348.

809. Garty, R., Bar-Schany, S. and Gordin, A. Possible transplacental transmission of serum hepatitis. Lancet, 1971, 2:434.

810. Gazala, E., Karplus, M., Liberman, J.R. and Sarov, I. The effect of maternal measles on the fetus. Pediatr. Infect. Dis., 1985, 4:203-204.

811. Gebreel, A.O. and Dane, D.S. Hepatitis in pregnancy in Libya. Ann. Trop. Med. Parasitol., 1983, 77:321-322.

812. Geeves, R.B., Lindsay, D.A. and Robertson, T.I. Varicella pneumonia in pregnancy with varicella neonatorum: report of a case followed by severe digital clubbing. Aust. N.Z. J. Med., 1971, 1:63-68.

813. Gehrz, R.C. Immunology and the host response to viral infection. Clin. Obstet. Gynecol., 1982, 25:545-554.

814. Gehrz, R.C. Neonatal herpes virus infections. Prevention, diagnosis and treatment. Part 1. Minn. Med., 1984, 67:379-380.

815. Gehrz, R.C., Marker, S.C., Knorr, S.O., Kalis, J.M. and Balfour, H.H., Jr. Specific cell-mediated immune defect in active cytomegalovirus infection of young children and their mothers. Lancet, 1977, 2:844-847.

816. Gelpi, A.P. Viral hepatitis complicating pregnancy: mortality trends in Saudi Arabia. Int. J. Gynaecol. Obstet., 1978, 17:73-77.

817. Geltzer, A.I., Guber, D. and Sears, M.L. Ocular manifestations of the 1964-65 rubella epidemic. Am. J. Ophthalmol., 1967, 63:221-229.

818. Geme, J.W., Jr. Alternate immunoprophylactic and biologic techniques in the control of viral respiratory tract diseases. Calif. Med., 1973, 118:6-11.

819. Gerety, R.J. Maternal transmission of hepatitis B. N. Engl. J. Med., 1976, 295:170-171.

820. Gerety, R.J. and Schweitzer, I.L. Viral hepatitis type B during pregnancy, the neonatal period, and infancy. J. Pediatr., 1977, 90:368-374.

821. German, L.J., McCracken, A.W. and Wilkie, K.M. Outbreak of febrile illness associated with E.C.H.O. virus type 5 in maternity unit in Singapore. Br. Med. J., 1968, 1:742-744.

822. Gershon, A., Fish, I. and Brunell, P. Herpes simplex infection of the newborn. Amer. J. Dis. Child, 1972, 124:739.

823. Gershon, A.A. Varicella in mother and infant: problems old and new. In: Krugman, S., Gershon, A.A., ed. Infections of the fetus and the newborn infant. New York, Liss, 1975, 79-95.

824. Gershon, A.A., Raker, R., Steinberg, S., Topf-Olstein, B. and Drusin, L.M. Antibody to Varicella-Zoster virus in parturient women and their offspring during the first year of life. Pediatrics, 1976, 58:692-696.

825. Gersony, W.M., Katz, S.L. and Nadas, A.S. Endocardial fibroelastosis and mumps virus. Pediatrics, 1966, 37:430.

826. Ghobrial, F.G. Pathogenesis of rubella cataract. Bull. Ophthalmol. Soc. Egypt, 1967, 60:301-302.

827. Gibbons, D.F., Winkler, C.E., Shaw, I.G., Terlecki, S., Richardson, C. and Done, J.T. Pathogenicity of the border disease agent for the bovine foetus. Br. Vet. J., 1974, 130:357-360.

828. Gibbs, C.E. Rubella immunization in women: test before shooting. Am. Fam. Physician, 1973, 8:145-147.

829. Gibbs, C.J., Jr., Gajdusek, D.C. and Alpers, M.P. Attempts to transmit subacute and chronic neurological diseases to animals. Int. Arch. Allergy Appl. Immunol., 1969, 36:519-552.

830. Gibbs, E.P. and Collings, D.F. Observations on bovine herpes mammillitis (BHM) virus infections of heavily pregnant heifers and young calves. Vet. Rec., 1972, 90:66-68.

831. Gibbs, E.P., Lawman, M.J. and Herniman, K.A. Preliminary observations on transplacental infection of bluetongue virus in sheep - a possible overwintering mechanism. Res. Vet. Sci., 1979, 27:118-120.

832. Gibbs, R.C., Shapiro, E., Cassidy, H. and Brunell, P.A. Possible mechanisms for maintaining immunity to varicella-zoster virus. Clinical observations. Am. J. Dis. Child, 1970, 120:456-457.

833. Gibbs, R.S. Infection control of herpes simplex virus infections in obstetrics and gynecology. J. Reprod. Med., 1986, 31:396-398.

834. Gibson, P.E., Field, A.M., Gardner, S.D. and Coleman, D.V. Occurrence of IgM antibodies against BK and JC polyomaviruses during pregnancy. J. Clin. Pathol., 1981, 34:674-679.

835. Gilbert, G.L. Congenital and perinatal cytomegalovirus infections. Aust. N.Z. J. Obstet. Gynaecol., 1985, 25:169-171.

836. Gilder, S.S. Rubella and pregnancy. Can. Med. Assoc. J., 1968, 99:726.

837. Giles, J.P., Cooper, L.Z. and Krugman, S. The rubella syndrome. J. Pediat., 1965, 66:434.

838. Giles, P.F., Mackay-Scollay, E.M. and Christiansen, K. Rubella antibody titres, rubella vaccination, and the obstetrician. Aust. N.Z. J. Obstet. Gynaecol., 1973, 13:65-71.

839. Gillespie, A., Dorman, D., Walker-Smith, J.A. and Yu, J.S. Neonatal hepatitis and Australia antigen. Lancet, 1970, 2:1081.

840. Gillick, J.C. An outbreak of swine foetal mummification associated with porcine parvovirus. Aust. Vet. J., 1977, 53:105-106.

841. Girardi, A.J., Reppucci, P., Dierlam, P., Rutala, W. and Coggin, J.H., Jr. Prevention of simian virus 40 tumors by hamster fetal tissue: influence of parity status of donor females on immunogenicity of fetal tissue and on immune cell cytotoxicity. Proc. Natl. Acad. Sci. U.S.A., 1973, 70:183-186.

842. Giraud, P., Drouet, J. and Dupuy, J.M. Hepatitis-B virus infection of children born to mothers with severe hepatitis. Lancet, 1975, 2:1088-1089.

843. Gladstone, J.L. and Millian, S.J. Rubella exposure in an obstetric clinic. Obstet. Gynecol., 1981, 57:182-186.

844. Glasser, L. and Delta, B.G. Congenital toxoplasmosis with placental infection in monozygotic twins. Pediatrics, 1965, 35:276.

845. Gledhill, B.L. Viral infertility in cattle. Cornell Vet., 1968, 58:466-479.

846. Gleeson, L.J. and Coggins, L. Response of pregnant mares to equine herpesvirus 1 (EHV 1). Cornell Vet., 1980, 70:391-400.

847. Gleeson, L.J. and Studdert, M.J. Equine herpesviruses. Experimental infection of a foetus with type 2. Aust. Vet. J., 1977, 53:360-362.

848. Gleeson, L.J., Sullivan, N.D. and Studdert, M.J. Equine herpesviruses: type 3 as an abortigenic agent. Aust. Vet. J., 1976, 52:349-354.

849. Glezen, W.P. Pathogenesis of bronchiolitis -- epidemiologic considerations. Pediatr. Res., 1977, 11:239-243.

850. Glover, S.G., Smith, C.C., Engeset, J. and Koruth, N.M. Unusual presentation of giant splenic artery aneurysm. Br. J. Surg., 1982, 69:247.

851. Gold, E. Viral infection of the fetus. West J. Med., 1974, 120:407-408.

852. Goldberg, G.N., Fulginiti, V.A., Ray, C.G., Ferry, P., Jones, J.F., Cross, H. and Minnich, L. In utero Epstein-Barr virus (infectious mononucleosis) infection. J.A.M.A., 1981, 246:1579-1581.

853. Goldsmith, M.F. Possible herpesvirus role in abortion studied. J.A.M.A., 1984, 251:3067-3070.

854. Goldwater, P.N., Quiney, J.R. and Banatvala, J.E. Maternal rubella at St. Thomas' Hospital: is there a need to change British vaccination policy? Lancet, 1978, 2:1298-1300.

855. Gomez, M.M. and Boxaca, M.C. Infection of pregnant guinea pigs with attenuated Junin virus strains. Intervirology, 1985, 23:8-14.

856. Gonik, B., Loo, L.S., West, S. and Kohl, S. Natural killer cell cytotoxicity and antibody-dependent cellular cytotoxicity to herpes simplex virus-infected cells in human pregnancy. Am. J. Reprod. Immunol. Microbiol., 1987, 13:23-26.

857. Goodpasture, E. Virus infection of the mammalian fetus. Science, 1942, 95:391.

858. Goodpasture, E.W. and Talbot, F.B. Concerning the nature of 'protozoan-like' cells in certain lesions in infancy. Am. J. Dis. Child, 1921, 21:415.

859. Goscienski, P.J. and Sexton, R.R. Follow-up studies in neonatal inclusion conjunctivitis. Am. J. Dis. Child, 1972, 124:180-182.

860. Goudeau, A., Lo, K.J., Coursaget, P., Tong, M.J., Yeh, C.L., Tsai, Y.T., Lee, J.K., Wu, T.C., Yeh, S.H. and Lee, S.D. Prevention of hepatitis B virus infection in children born to HBsAg positive/HBeAg positive mothers. Preliminary results of active and passive-active immunization. Dev. Biol. Stand., 1983, 54:399-404.

861. Goudeau, A., Yvonnet, B., Lesage, G., Barin, F., Denis, F., Coursaget, P., Chiron, J.P. and Diop-Mar, I. Lack of anti-HBc IgM in neonates with HBsAg carrier mothers argues against transplacental transmission of hepatitis B virus infection. Lancet, 1983, 2:1103-1104.

862. Gould, J.J. and Mims, C.A. Murine cytomegalovirus: reactivation in pregnancy. J. Gen. Virol., 1980, 51:397-400.

863. Gourlay, J.A. Comments on the use of measles virus for distemper prophylaxis. J. Am. Vet. Med. Assoc., 1970, 156:1769-1770.

864. Goyert, G.L., Bottoms, S.F. and Sokol, R.J. Anicteric presentation of fatal herpetic hepatitis in pregnancy. Obstet. Gynecol., 1985, 65:585-588.

865. Goyette, R.E., Donowho, E.M., Jr., Hieger, L.R. and Plunkett, G.D. Fulminant herpesvirus hominis hepatitis during pregnancy. Obstet. Gynecol., 1974, 43:191-195.

866. Graber, E.A., Barber, H.R. and ORourke, J.J. Simple surgical treatment for condyloma acuminatum of the vulva. Obstet. Gynecol., 1967, 29:247-250.

867. Granat, M., Morag, A., Margalioth, E.J., Leviner, E. and Ornoy, A. Fetal outcome following primary herpetic gingivostomatitis in early pregnancy. Morphological study and updated appraisal. Isr. J. Med. Sci., 1986, 22:455-459.

868. Grangeot-Keros, L., Nicolas, J.C., Bricout, F. and Pillot, J. Rubella reinfection and the fetus. N. Engl. J. Med., 1985, 313:1547.

869. Grant, S., Edmond, E. and Syme, J. A prospective study of cytomegalovirus infection in pregnancy. I. Laboratory evidence of congenital infection following maternal primary and reactivated infection. J. Infect., 1981, 3:24-31.

870. Gray, E.S., Anand, A. and Brown, T. Parvovirus infections in pregnancy. Lancet, 1986, 1:208.

871. Green, D.K., Reid, .M. and Rhaney, K. Generalised vaccinia in the human foetus. Lancet, 1966, 1296.

872. Green, D.M., Reid, S.M. and Rhaney, K. Generalised vaccinia in the human foetus. Lancet, 1966, 1:1296-1298.

873. Green, R.H., Balsamo, M.R., Giles, J.P., Krugman, S. and Mirick, G.S. Studies of the natural history and prevention of rubella. Am. J. Dis. Child, 1965, 110:348-365.

874. Greenberg, M., Jacobziner, H., Pakter, J. and Weisl, B.A.G. Maternal mortality in epidermic Asian influenza. Am. J. Obstet. Gynecol., 1958, 76:897.

875. Greenberg, M., Yankauer, A. Jr. and Krugman, S. et al. Effect of smallpox vaccination during pregnancy on the incidence of congenital malformations. Pediatrics, 1949, 3:456.

876. Greenberg, M.W. and Beilly, J.S. Congenital defects in the infant following mumps during pregnancy. Amer. J. Obst. Gyn., 1949, 57:805.

877. Greene, C.E., Gorgasz, E.J. and Martin, C.L. Hydraencephaly associated with feline panleukopenia. J. Am. Vet. Med. Assoc., 1982, 180:767-768.

878. Greenwood, R.D. Intrauterine infection, 1808. J. Okla. State Med. Assoc., 1975, 68:416-417.

879. Greffe, B.S., Dooley, S.L., Deddish, R.B. and Krasny, H.C. Transplacental passage of acyclovir. J. Pediatr., 1986, 108:1020-1021.

880. Gregg, N.M. Congenital cataract following German measles in the mother. Trans. Ophthal. Soc. Aust., 1941, 3:35.

881. Gregg, N.M. Further observations on congenital defects in infants following maternal rubella. Trans. Ophthalmol. Soc. Aust., 1944, 4:119-131.

882. Gregg, N.M., Beavis, W.R., Heseltine, M., Macklin, A.E. and Vickery, D. Occurrence of congenital defects in children following maternal rubella during pregnancy. Med. J. Aust., 1945, 2:122-126.

883. Gregoriadis, G. Targeting of drugs: possibilities in viral chemotherapy and prophylaxis. Pharmacol. Ther. 1980, 10: 103-118.

884. Griffith, B.P. and Hsiung, G.D. Cytomegalovirus infection in guinea pigs. IV. Maternal infection at different stages of gestation. J. Infect. Dis., 1980, 141:787-793.

885. Griffith, B.P., Lavallee, J.T., Jennings, T.A. and Hsiung, G.D. Transmission of maternal cytomegalovirus-specific immunity in the guinea pig. Clin. Immunol. Immunopathol., 1985, 35:169-181.

886. Griffith, B.P., Lucia, H.L., Tillbrook, J.L. and Hsiung, G.D. Enhancement of cytomegalovirus infection during pregnancy in guinea pig. J. Infect. Dis., 1983, 147:990-998.

887. Griffith, B.P., McCormick, S.R., Booss, J. and Hsiung, G.D. Inbred guinea pig model of intrauterine infection with cytomegalovirus. Am. J. Pathol., 1986, 122:112-119.

888. Griffith, B.P., McCormick, S.R., Fong, C.K., Lavallee, J.T., Lucia, H.L. and Goff, E. The placenta as a site of cytomegalovirus infection in guinea pigs. J. Virol., 1985, 55:402-409.

889. Griffith, E.R. Viral pneumonia in pregnancy: report of a case complicated by disseminated intravascular coagulation and acute renal failure. Am. J. Obstet. Gynecol., 1974, 120:201-202.

890. Griffith, G.W., Adelstein, A.M., Lambert, P.M. and Weatherall, J.A. Influenza and infant mortality. Br. Med. J., 1972, 3:553-556.

891. Griffith, J.F. Nonbacterial infections of the fetus and newborn. Clin. Perinatol., 1977, 4:117-130.

892. Griffiths, P., Baboonian, C. and Ashby, D. The demographic characteristics of pregnant women infected with cytomegalovirus. Int. J. Epidemiol., 1985, 14:447-452.

893. Griffiths, P.D. The presumptive diagnosis of primary cytomegalovirus infection in early pregnancy by means of a radioimmunoassay for specific - IgM antibodies. Br. J. Obstet. Gynaecol., 1981, 88:582-587.

894. Griffiths, P.D. Cytomegalovirus and the liver. Semin. Liver Dis., 1984, 4:307-313.

895. Griffiths, P.D. and Baboonian, C. A prospective study of primary cytomegalovirus infection during pregnancy: final report. Br. J. Obstet. Gynaecol., 1984, 91:307-315.

896. Griffiths, P.D. and Baboonian, C. Intra-uterine transmission of cytomegalovirus in women known to be immune before conception. J. Hyg. (Lond.), 1984, 92:89-95.

897. Griffiths, P.D., Campbell-Benzie, A. and Health, R.B. A prospective study of primary cytomegalovirus infection in pregnant women. Br. J. Obstet. Gynaecol., 1980, 87:308-314.

898. Griffiths, P.D., Ronald, C.J. and Heath, R.B. A prospective study of influenza infections during pregnancy. J. Epidemiol. Community Health, 1980, 34:124-128.

899. Griffiths, P.D., Stagno, S., Pass, R.F., Smith, R.J. and Alford, C.A., Jr. Congenital cytomegalovirus infection: diagnostic and prognostic significance of the detection of specific immunoglobulin M antibodies in cord serum. Pediatrics, 1982, 69:544-549.

900. Griffiths, P.D., Stagno, S., Pass, R.F., Smith, R.J. and Alford, C.A., Jr. Infection with cytomegalovirus during pregnancy: specific IgM antibodies as a marker of recent primary infection. J. Infect. Dis., 1982, 145:647-653.

901. Grillner, L., Forsgren, M., Barr, B., Bottiger, M., Danielsson, L. and DeVerdier, C. Outcome of rubella during pregnancy with special reference to the 17th - 24th weeks of gestation. Scand. J. Infect. Dis., 1983, 15:321-325.

902. Gringras, M. and Bowker, C. Rubella susceptibility in women. Br. Med. J., 1978, 2:698.

903. Grist, N.R. and Bell, E.J. A six year study of Coxsackievirus B infections in heart disease. J. Hyg. Cambridge, 1974, 73:165.

904. Gronroos, M., Honkonen, E., Terho, P. and Punnonen, R. Cervical and serum IgA and serum IgG antibodies to Chlamydia trachomatis and herpes simplex virus in threatened abortion: a prospective study. Br. J. Obstet. Gynaecol., 1983, 90:167-170.

905. Gronvall, H. and Selander, P. Some virus diseases during pregnancy and their effect on the fetus. Nord. Med., 1948, 37:409.

906. Gross, L. The role of viruses in the etiology of cancer and leukemia. J.A.M.A., 1974, 230:1029-1032.

907. Grossman, J.H., III. Perinatal viral infections. Clin. Perinatol., 1980, 7:257-271.

908. Grossman, J.H., III. Viral infections in obstetrics and gynecology. Obstet. Gynecol. Annu., 1980, 9:55-76.

909. Grossman, J.H., III. Herpes simplex virus (HSV) infections. Clin. Obstet. Gynecol., 1982, 25:555-561.

910. Grossman, J.H., III. Preventing neonatal herpes simplex infections: an obstetrician's perspective. Prog. Clin. Biol. Res., 1985, 163B:351-354.

911. Grossman, J.H., III, Wallen, W.C. and Sever, J.L. Management of genital herpes simplex virus infection during pregnancy. Obstet. Gynecol., 1981, 58:1-4.

912. Grover, L., Kane, J., Kravitz, J. and Cruz, A. Systemic acyclovir in pregnancy: a case report. Obstet. Gynecol., 1985, 65:284-287.

913. Growdon, W.A., Apodaca, L., Cragun, J., Peterson, E.M. and DeLaMaza, L.M. Neonatal herpes simplex virus infection occurring in second twin of as asymptomatic mother. Failure of a modern protocol. J.A.M.A., 1987, 257:508-511.

914. Guggenheimer, J., Nowak, A.J. and Michaels, R.H. Dental manifestations of the rubella syndrome. Oral Surg. Oral Med. Oral Pathol., 1971, 32:30-37.

915. Guha, D.K., Rashmi, A. and Khanduja, P.C. In utero infection of the fetus by herpes simplex virus. Indian J. Pediatr., 1968, 35:484-485.

916. Guglielmino, S., Tempera, G., Pappalardo, G. and Castro, A. H-1 and X14 parvovirus antibodies in women with abortions or still-births. Acta. Virol. (Praha.), 1978, 22:426-428.

917. Gullikson, J.S. Tooth morphology in rubella syndrome children. J. Dent. Child, 1975, 42:479-482.

918. Gumpel, S.M. Clinical and social status of patients with congenital rubella. Arch. Dis. Child, 1972, 47:330-337.

919. Gumpel, S.M., Hayes, K. and Dudgeon, J.A. Congenital perceptive deafness: role of intrauterine rubella. Br. Med. J., 1971, 2:300-304.

920. Gunby, P. California battles perinatal hepatitis B. J.A.M.A., 1982, 247:1238.

921. Gunther, M. Cot deaths: anaphylactic reaction after intrauterine infection as another potential cause. Lancet, 1966, 1:912-914.

922. Gupta, B., Agarwal, S. and Joshi, V.V. Maternal/fetal transmission of HBsAg negative hepatitis. Lancet, 1978, 2:740.

923. Gupta, J.D. and Murphy, A.M. Rubella-specific IgM antibody as an aid to the diagnosis of acquired and congenital rubella. Am. Heart J., 1975, 89:258-259.

924. Gupta, M.L., Sharma, B.S., Sharma, U., Saxena, S., Sharma, M.L. and Sharma, J.N. Hepatitis-B surface antigen in pregnant women and their offsprings. Indian Pediatr., 1986, 23:47-50.

925. Gupta, M.L., Sharma, N.J., Sharma, B.S., Sharma, U., Saxena, S. and Sharma, M.L. Maternal and fetal outcome of viral hepatitis during pregnancy. Indian Pediatr., 1986, 23:911-914.

926. Gupta, M.L., Sharma, U., Saxena, S., Sharma, M.L. and Pokharna, D.S. Vertical transmission of hepatitis-B surface antigen from asymptomatic carrier mothers. Indian Pediatr., 1985, 22:339-342.

927. Gupta, P. and Ferrer, J.F. Comparison of various serological and direct methods for the diagnosis of BLV infection in cattle. Int. J. Cancer, 1981, 28:179-184.

928. Gurwith, M., Wenman, W., Hinde, D., Feltham, S. and Greenberg, H. A prospective study of rotavirus infection in infants and young children. J. Infect. Dis., 1981, 144:218-224.

929. Gussetti, N., Pornaro, E., Largajolli, G. and Delia, R. Vertical transmisssion of HBV from mothers HBsAg positive, anti-HBe positive. Dev. Biol. Stand., 1983, 54:405-408.

930. Guy, J.S. and Potgieter, L.N. Bovine herpesvirus-1 infection of cattle: kinetics of antibody formation after intranasal exposure and abortion induced by the virus. Am. J. Vet. Res., 1985, 46:893-898.

# H

931. Haddad, J., Simeoni, U., Messer, J. and Willard, D. Perinatal varicella. Lancet, 1986, 1:1494-1495.

932. Hadji, G., Meszner, Z., Nyerges, G. et al. Congenital varicella syndrome. Infection, 1986, 14:177-180.

933. Hadjisavvas, T.H., Harkness, J.W., Huck, R.A. and Stuart, P. The demonstration by interference tests of an infective agent in fetuses from ewes inoculated with Border disease tissue. Res. Vet. Sci., 1975, 18:237-243.

934. Hadlow, W.J., Jackson, T.A. and Race, R.E. Experimental infection of fetal and newborn Suffolk sheep with scrapie virus. Am. J. Vet. Res., 1984, 45:2637-2639.

935. Haemmerli, U.P. Jaundice during pregnancy. Acta Med. Scand., 1966, 444:17.

936. Haggerty, L. TORCH: a literature review and implications for practice. J. Obstet. Gynecol. Neonatal Nurs., 1985, 14:124-129.

937. Hagler, W.S., Walters, P.V. and Nahmias, A.J. Ocular involvement in neonatal herpes simplex virus infection. Arch. Ophthalmol., 1969, 82:169-176.

938. Hahimi, A., Carruthers, M.M., Wolf, P. and Lerner, A.M. Congenital infections with reovirus. J. Exp. Med., 1966, 124:33-46.

939. Hain, J., Doshi, N. and Harger, J.H. Ascending transcervical herpes simplex infection with intact fetal membranes. Obstet. Gynecol., 1980, 56:106-109.

940. Hajek, P., Kovaru, F. and Kruml, J. The antibody response and morphological changes following immunization of pig foetuses with phi chi 174 phage. Folia. Microbiol. (Praha.), 1969, 14:492-495.

941. Hakosalo, J. and Saxen, L. Influenza epidemic and congenital defects. Lancet, 1971, 2:1346-1347.

942. Hakulinen, T., Hovi, L., Karkinen-Jaaskelainen, Penttinen, K. and Saxen, L. Association between influenza during pregnancy and childhood leukaemia. Br. Med. J., 1973, 4:265-267.

943. Haldane, E.V., Van-Rooyen, C.E., Embil, J.A., Tupper, W.C., Gordon, P.C. and Wanklin, J.M. A search for transmissible birth defects of virologic origin in members of the nursing profession. Am. J. Obstet. Gynecol., 1969, 105:1032-1040.

944. Hall, E.G., Hay, J.D., Moss, P.D. and Ryan, M.M.P. Congenital toxoplasmosis in the newborn. Arch. Dis. Child, 1953, 28:117-124.

945. Hall, J.A., Wills, E.R., Hall, K. and Pearson, J.W. Positive rubella titers in pregnancy: a question of validity. Obstet. Gynecol., 1979, 54:235-236.

946. Halstead, E., Halstead, S.B., Jackson, R.S., Char, D., Hale, R. and Pion, R. Rubella vaccination: fertility control in a large-scale vaccination program for postpubertal women. Am. J. Obstet. Gynecol., 1975, 121:1089-1094.

947. Hamashima, Y. Intrauterine and neonatal infections under immature antibody formation system. Nippon. Ketsueki. Gakkai. Zasshi., 1967, 30:104-114.

948. Hamazaki, M. Histological study of congenital and acquired cytomegalovirus infection. Acta. Pathol. Jpn., 1983, 33:89-96.

949. Hamberger, V. and Habel, K. Teratogenic and lethal effects of influenza A and mumps viruses on early chick embryos. Proc. Soc. Exp. Biol. Med., 1947, 66:608.

950. Hambling, M.H. Changes in the distribution of rubella antibodies in women of childbearing age during the first eight years of a rubella vaccination programme. J. Infect., 1980, 2:341-346.

951. Hammond, G.W. Neonatal herpes simplex. Can. Med. Assoc. J., 1982, 126:468-469.

952. Hammond, G.W., Lukes, H., Wells, B., Thompson, L., Low, D.E. and Cheang, M. Maternal and neonatal neutralizing antibody titers to selected enteroviruses. Pediatr. Infect. Dis., 1985, 4:32-35.

953. Hammond, K. Skin dimples and rubella. Pediatrics, 1967, 39:291-292.

954. Hancock, M.P., Huntley, C.C. and Sever, J.L. Congenital rubella syndrome with immunoglobulin disorder. J. Pediat., 1968, 72:636.

955. Hanid, T.K. Hypothyroidism in congenital rubella. Lancet, 1976, 2:854.

956. Hankins, G.D., Cunningham, F.G., Luby, J.P., Butler, S.L., Stroud, J. and Roark, M. Asymptomatic genital excretion of herpes simplex virus during early labor. Am. J. Obstet. Gynecol., 1984, 150:100-101.

957. Hann, H.W., Kim, C.Y., London, W.T., Whitford, P. and Blumberg, B.S. Hepatitis B virus and primary hepatocellular carcinoma: family studies in Korea. Int. J. Cancer, 1982, 30:47-51.

958. Hanna, C. Embryopathic time-table of cataract formation in congenital rubella. J. Pediatr. Ophthalmol., 1976, 13:266-273.

959. Hanna, M.G., Jr., Tennant, R.W. and Coggin, J.H., Jr. Suppressive effect of immunization with mouse fetal antigens on growth of cells infected with Rauscher leukemia virus and plasma-cell tumors. Proc. Natl. Acad. Sci. U.S.A., 1971, 68:1748-1752.

960. Hanngren, K., Grandien, M. and Granstrom, G. Effect of zoster immunoglobulin for varicella prophylaxis in the newborn. Scand. J. Infect. Dis., 1985, 17:343-347.

961. Hannon, W.H., Ambrose, K.R. and Coggin, J.H., Jr. Forssman antigen and phase specific fetal antigens: an evaluation of their role in SV40 tumor immunity. Proc. Soc. Exp. Biol. Med., 1975, 148:1075-1080.

962. Hanshaw, J.B. Congenital and acquired cytomegalovirus infection. Pediat. Clin. N. Amer., 1966, 13:279.

963. Hanshaw, J.B. Cytomegalovirus complement fixing antibody in microcephaly. N. Engl. J. Med., 1966, 275:476-479.

964. Hanshaw, J.B. Congenital cytomegalovirus infection: laboratory methods of detection. J. Pediatr., 1969, 75:1179-1185.

965. Hanshaw, J.B. Developmental abnormalities associated with congenital cytomegalovirus infection. Adv. Teratol., 1970, 4:62-93.

966. Hanshaw, J.B. Congenital cytomegalovirus infection: A fifteen year perspective. J. Infect. Dis., 1971, 123:555-561.

967. Hanshaw, J.B. Congenital cytomegalovirus infection. N. Engl. J. Med., 1973, 288:1406-1407.

968. Hanshaw, J.B. Herpesvirus hominis infections in the fetus and the newborn. Am. J. Dis. Child, 1973, 126:546-555.

969. Hanshaw, J.B. On deafness, cytomegalovirus, and neonatal screening. Am. J. Dis. Child, 1982a, 136:886-887.

970. Hanshaw, J.B. Perinatal infections: prevention of long-term sequelae. Ciba Found. Symp., 1979, 77:247-260.

971. Hanshaw, J.B. and Dudgeon, J.A. Congenital cytomegalovirus. Major Probl. Clin. Pediatr., 1978, 17:97-152.

972. Hanshaw, J.B. and Dudgeon, J.A. Enteroviral infections. Major Probl. Clin. Pediatr., 1978, 17:182-191.

973. Hanshaw, J.B. and Dudgeon, J.A. Evidence for the viral etiology of congenital defects. Major Probl. Clin. Pediatr., 1978, 17:10-16.

974. Hanshaw, J.B. and Dudgeon, J.A. Hepatitis viruses. Major Probl. Clin. Pediatr., 1978, 17:222-229.

975. Hanshaw, J.B. and Dudgeon, J.A. Herpes simplex infection of the fetus and newborn. Major Probl. Clin. Pediatr., 1978, 17:153-181.

976. Hanshaw, J.B. and Dudgeon, J.A. Laboratory diagnosis. Major Probl. Clin. Pediatr., 1978, 17:301-318.

977. Hanshaw, J.B. and Dudgeon, J.A. Other possible pathogens in the fetus and newborn. Major Probl. Clin. Pediatr., 1978, 17:230-236.

978. Hanshaw, J.B. and Dudgeon, J.A. Prevention, treatment, and antiviral chemotherapy. Major Probl. Clin. Pediatr., 1978, 17:327-347.

979. Hanshaw, J.B. and Dudgeon, J.A. Rubella. Major Probl. Clin. Pediatr., 1978, 17:17-96.

980. Hanshaw, J.B. and Dudgeon, J.A. Smallpox and vaccinia. Major Probl. Clin. Pediatr., 1978, 17:209-221.

981. Hanshaw, J.B. and Dudgeon, J.A. Varicella-zoster infections. Major Probl. Clin. Pediatr., 1978, 17:192-208.

982. Hanshaw, J.B. and Dudgeon, J.A. Viral diseases of the fetus and newborn. Introduction. Major Probl. Clin. Pediatr., 1978, 17:1-9.

983. Hanshaw, J.B., Scheiner, A.P., Moxley, A.W., Gaev, L., Abel, V. and Scheiner, B. School failure and deafness after "silent" congenital cytomegalovirus infection. N. Engl. J. Med., 1976, 295:468-470.

984. Hanshaw, J.B., Schultx, F.W., Melish, M.M. and Dudgeon, J.A. Congenital cytomegalovirus infections. Ciba. Found. Symp., 1972, 10:23-43.

985. Hanson, L.A., Lundgren, S., Lycke, E., Strannegard, O. and Winberg, J. Clinical and serological observations in cases of Coxsackie B3 infections in early infancy. Acta. Paediatr. Scand., 1966, 55:577-583.

986. Hansson, O., Kristensson, K., Lycke, E., Solymar, L. and Sourander, P. Generalized myopathy and cerebral malformations possibly related to an enteroviral infection. Acta. Paediatr. Scand., 1975, 64:881-885.

987. Happel, J.S. Mastitis in the male -- a rare complication of mumps. Br. Med. J., 1965, 1041.

988. Hardin, J.A. Cyclophosphamide treatment of lymphoma during third trimester of pregnancy. Obstet. Gynecol., 1972, 39:850-851.

989. Harding, J.D., Done, J.T. and Darbyshire, J.H. Congenital tremors in piglets and their relation to swine fever. Vet. Rec., 1966, 79:388-390.

990. Hardy, J.B. Viral infection in pregnancy: a review. Am. J. Obstet. Gynecol., 1965, 93:1052-1065.

991. Hardy, J.B. Viruses and the fetus. Postgrad. Med., 1968, 43:156-165.

992. Hardy, J.B. Rubella as a teratogen. Birth Defects, 1971, 7:64-71.

993. Hardy, J.B. Clinical and developmental aspects of congenital rubella. Arch. Otolaryngol., 1973, 98:230-236.

994. Hardy, J.B. Fetal consequences of maternal viral infections in pregnancy. Arch. Otolaryngol., 1973, 98:218-227.

995. Hardy, J.B. Immediate and long-range effects of maternal viral infection in pregnancy. Birth Defects, 1976, 12:23-31.

996. Hardy, J.B., McCracken, G.H., Jr., Gilkeson, M.R. and Sever, J.L. Adverse fetal outcome following maternal rubella after the first trimester of pregnancy. J.A.M.A., 1969, 207:2414-2420.

997. Hardy, W.D., Jr., Hess, P.W., Essex, M., Cotter, S., McClelland, A.J. and MacEwen, G. Horizontal transmission of feline leukemia virus in cats. Bibl. Haematol., 1975, 67-74.

998. Hare, J. Neonatal encephalitis following maternal genital herpesvirus infection. Proc. R. Soc. Med., 1974, 67:15-16.

999. Harger, J.H. Herpes simplex encephalitis in pregnancy. Am. J. Obstet. Gynecol., 1980, 137:520.

1000. Harger, J.H., Meyer, M.P. and Amortegui, A.J. Changes in the frequency of genital herpes recurrences as a function of time. Obstet. Gynecol., 1986, 67:637-642.

1001. Harger, J.H., Pazin, G.J., Armstrong, J.A., Breinig, M.C. and Ho, M. Characteristics and management of pregnancy in women with genital herpes simplex virus infection. Am. J. Obstet. Gynecol., 1983, 145:784-791.

1002. Harley, J.D. Single-sample diagnosis of recent rubella. Med. J. Aust., 1971, 1:292.

1003. Harley, J.D. The scourge of rubella. Med. J. Aust., 1971, 2:457-459.

1004. Harley, J.D. and Gillespie, A.M. A complicated case of congenital vaccinia. Pediatrics, 1972, 50:150-153.

1005. Harper, P.S. Mendelian inheritance or transmissible agent? The lesson Kuru and the Australia antigen. J. Med. Genet., 1977, 14:389-398.

1006. Harrington, D.G., Lupton, H.W., Crabbs, C.L., Peters, C.J., Reynolds, J.A. and Slone, T.W., Jr. Evaluation of a formalin-inactivated Rift Valley fever vaccine in sheep. Am. J. Vet. Res., 1980, 41:1559-1564.

1007. Harris, R.E. Viral teratogenesis: a review with experimental and clinical perspectives. Am. J. Obstet. Gynecol., 1974, 119:996-1008.

1008. Harris, R.E., Morahan, P. and Coleman, P. Teratogenic effects of Colorado tick fever virus in mice. J. Infect. Dis., 1975, 131:397-402.

1009. Harris, W.H. Hepatitis B in pregnancy. Can. Med. Assoc. J., 1974, 110:1133.

1010. Hashiguchi, Y., Nanba, K. and Kumagai, T. Congenital abnormalities in newborn lambs following Akabane virus infection in pregnant ewes. Natl. Inst. Anim. Health Q (Tokyo), 1979, 19:1-11.

1011. Hashimoto, A., Hirai, K., Okada, K. and Fujimoto, Y. Pathology of the placenta and newborn pups with suspected intrauterine infection of canine herpesvirus. Am. J. Vet. Res., 1979, 40:1236-1240.

1012. Hashimoto, A., Hirai, K., Suzuki, Y. and Fujimoto, Y. Experimental transplacental transmission of canine herpesvirus in pregnant bitches during the second trimester of gestation. Am. J. Vet. Res., 1983, 44:610-614.

1013. Hashimoto, A., Hirai, K., Yamaguchi, T. and Fujimoto, Y. Experimental transplacental infection of pregnant dogs with canine herpesvirus. Am. J. Vet. Res., 1982, 43:844-850.

1014. Hassan, S.A. and Cochran, K.W. Effects of reovirus type 1 on the developing mouse embryo. Am. J. Pathol., 1969, 55:147-161.

1015. Hastreiter, A.R., Joorabchi, B., Pujatti, G., VanDerHorst, R., Pataesil, G. and Sever, J.L. Cardiovascular lesions associated with congenital rubella. J. Pediatr., 1967, 71:59-65.

1016. Hatherley, L.I. The incidence of cytomegalic inclusion disease (CID) in an obstetric teaching hospital, 1975-1984. Aust. N.Z. J. Obstet. Gynaecol., 1985, 25:171-175.

1017. Hau, J. and Buschard, K. Effect of encephalomyocarditis (EMC) virus on murine foetal and placental growth monitored by quantification of maternal plasma levels of pregnancy-associated murine protein-2 and alpha-fetoprotein. Acta. Pathol. Microbiol. Immunol. Scand., 1986, 94:339-342.

1018. Haukenes, G., Finne, P.H., Bertnes, E., Orstavik, I., Tjtta, E., Haneberg, B. and Aarseth, P. Cytomegalovirus (CMV) and rubella virus infection during pregnancy. A study of CMV and rubella virus antibodies in 2014 pregnant women and follow-up studies of infants at risk for intrauterine CMV infections. Acta. Obstet. Gynecol. Scand., 1984, 63:431-435.

1019. Haukenes, G. and Haram, K.O. Clinical rubella after reinfection. N. Engl. J. Med., 1972, 287:1204.

1020. Haukenes, G., Haram, K. and Solberg, C.O. Rubella reinfection and the fetus. Lancet, 1973, 1:1313.

1021. Haveri, R. and Dosik, H. Neonatal hepatitis B. Compr. Ther., 1980, 6:59-63.

1022. Hayden, G.F., Herrmann, K.L., Buimovici-Klein, E., Weiss, K.E., Nieburg, P.I. and Mitchell, J.E. Subclinical congenital rubella infection associated with maternal rubella vaccination in early pregnancy. J. Pediatr., 1980, 96:869-872.

1023. Hayder, H.A., Storz, J. and Young, S. Antigenicity of bovine parvovirus in fetal infections. Am. J. Vet. Res., 1983, 44:558-563.

1024. Hayes, K. Prenatal viral infection with particular reference to cytomegaloviruses. Aust. Paediatr. J., 1974, 10:56-63.

1025. Hayes, K., Dudgeon, J.A. and Soothill, J.F. Humoral immunity in congenital rubella. Clin. Exp. Immunol., 1967, 2:653-657.

1026. Hayes, K. and Gibas, H. Placental cytomegalovirus infection without fetal involvement following primary infection in pregnancy. J. Pediatr., 1971, 79:401-405.

1027. Hayes, K., Symington, G. and Mackay, I.R. Maternal immunosuppression and cytomegalovirus infection of the fetus. Aust. N.Z. J. Med., 1979, 9:430-433.

1028. Haymaker, W., Girdany, B.R., Stephens, J., Lille, R.D. and Fetterman, G.H. Cerebral involvement with advanced periventricular calcification in generalized cytomegalic inclusion disease in the newborn. J. Neuropathol. Exp. Neurol., 1954, 13:562-586.

1029. Hazrati, A. and Dayhim, F. Serological survey on infection with equine rhinopneumonitis virus among soliped animals in Iran. Vet. Rec., 1973, 93:70-72.

1030. Heath, H.D., Shear, H.H., Imagawa, D.T., Jones, M.H. and Adams, J.M. Teratogenic effects of herpes simplex, vaccinia, influenza A (NWS), and distemper virus on early chick embryos. Proc. Soc. Exp. Biol. Med., 1956, 92:675.

1031. Heggie, A.D. Intrauterine infection in maternal rubella. J. Pediatr., 1967, 71:777-782.

1032. Heggie, A.D. Growth inhibition of human embryonic and fetal rat bones in organ culture by rubella virus. Teratology, 1977. 15:47.

1033. Heijtink, R.A., Backx, G., VanDerHorst, J.M. and Masurel, N. Complement fixation and neutralization RS antibodies in maternal and neonatal sera. J. Hyg. (Camb.), 1977, 78:411-417.

1034. Heijink, R.A., Boender, P.J., Schalm, S.W. and Masurel, N. Hepatitis B virus DNA in serum of pregnant women with HBsAg and HBeAg or antibodies to HBe. J. Infect. Dis., 1984, 150:462.

1035. Heineman, H.S. Infectious mononucleosis and fetal abnormality. Am. J. Obstet. Gynecol., 1979, 133:939.

1036. Heinonen, O.P., Shapiro, S., Monson, R.R., Hartz, S.C., Rosenberg, L. and Stone, D. Immunization during pregnancy against poliomyelitis and influenza in relation to childhood malignancy. Int. J. Epidemiol., 1973, 2:229-235.

1037. Helander, I., Arstila, P. and Terho, P. Herpes zoster in a 6-month-old infant. Acta. Derm. Venereol. (Stockh.), 1983, 63:180-181.

1038. Helgason, T. and Jonasson, M.R. Evidence for a food additive as a cause of ketosis-prone diabetes. Lancet, 1981, 2:716-720.

1039. Hemenway, W.G., Sando, I. and McChesney, D. Temporal bone pathology following maternal rubella. Arch. Klin. Exp. Ohren. Nasen. Kehlkopfheilkd., 1969, 193:287-300.

1040. Hendricks, A.G. Teratological findings in a baboon colony. FDA Conference on Non-human Primate Toxicology, Airlie House. Virginia, USA, 1966.

1041. Hensleigh, P.A., Glover, D.B. and Cannon, M. Systemic Herpesvirus hominis in pregnancy. J. Reprod. Med., 1979, 22:171-176.

1042. Herbert, L. and Rodger, J.A. Outbreak of equine herpesvirus abortion in the Hunter Valley, New South Wales. Equine Vet. J., 1983, 15:276-278.

1043. Herman, J., Sayar, Y., Gotlieb-Stematsky, T. Pregnancy following cytomegalovirus mononucleosis. Isr. J. Med. Sci., 1982, 18, 491-493.

1044. Herrmann, K.L. Congenital and perinatal varicella. Clin. Obstet. Gynecol., 1982, 25:605-609.

1045. Hermanns, W., Trautwein, G., Meyer, H. and Liess, B. Experimental transplacental transmission of hog cholera virus in pigs. V. Immunopathological findings in newborn pigs. Zentralbl. Veterinarmed. (B), 1981, 28:669-683.

1046. Hertzberg, R. Rubella and virus induced cataracts. Trans. Ophthalmol. Soc. U.K., 1982, 102:355-358.

1047. Hertzberg, R. Congenital cataract following German measles in the mother. Abstracts from the publications of the late Sir Norman McAlister Gregg. Aust. N.Z.J. Ophthalmol., 1985, 13:303-309.

1048. Hess, R.G. and Bachmann, P.A. Distribution of antibodies to rotavirus in serum and lacteal secretions of naturally infected swine and their suckling pigs. Am. J. Vet. Res., 1981, 42:1149-1152.

1049. Hibbard, L.T. Maternal/fetal bilirubin transfer. Am. J. Obstet. Gynecol., 1975, 121:282-283.

1050. Hicks, D.E. Comparison profiles of rubella and non-rubella deaf children. Am. Ann. Deaf., 1970, 115:86-92.

1051. Hieber, J.P., Dalton, D., Shorey, J. and Combes, B. Hepatitis and pregnancy. J. Pediatr., 1977, 91:545-549.

1052. Higa, K., Dan, K. and Manabe, H. Varicella-zoster virus infections during pregnancy: hypothesis concerning the mechanisms of congenital malformations. Obstet. Gynecol., 1987, 69:214-222.

1053. Higgins, I.T.T. The epidemiology of congenital heart disease. J. Chron. Dis., 1965, 18:699.

1054. Highman, J.H. Congenital osseous rubella. Clin. Radiol., 1967, 18:445-449.

1055. Hildebrandt, R.J., Sever, J.L., Margileth, A.M. and Callagan, D.A. Cytomegalovirus in the normal pregnant woman. Obst. Gyn., 1967, 98:1125-1128.

1056. Hillard, P., Seeds, J. and Cefalo, R. Disseminated herpes simplex in pregnancy: two cases and a review. Obstet. Gynecol. Surv., 1982, 37:449-453.

1057. Hilleman, M.R. Toward prophylaxis of prenatal infection by viruses. Obstet. Gynecol., 1969, 33:461-469.

1058. Hino, S., Yamaguchi, K., Katamine, S., Sugiyama, H., Amagasaki, T., Kinoshita, K., Yoshida, Y., Doi, H., Tsuji, Y. and Miyamoto, T. Mother-to-child transmission of human T-cell leukemia virus type-I. Gann., 1985, 76:474-480.

1059. Hirano, T., Miyajima, H., Watanabe, T., Tsukuda, R. and Shimamoto, K. Enhancement of tumor induction in rats with Moloney murine sarcoma virus by a "new" method based on direct injection into fetuses. J. Natl. Cancer Inst., 1977, 58:73-82.

1060. Hirayama, M. Rubella and congenital rubella syndrome in Japan. Paediatr. Univ. Tokyo, 1970, 18:41-44.

1061. Hitchins, A.J., Gostling, J.V., and ODriscoll, W.B. Serum hepatitis in late pregnancy. Lancet, 1974, 2:343-344.

1062. Ho-Yen, D.O., Hardie, R. and Sommerville, R.G. Varicella-induced thrombocytopenia. J. Infect., 1984, 8:274-276.

1063. Hockberger, R.S. and Rothstein, R.J. Varicella pneumonia in adults: a spectrum of disease. Ann. Emerg. Med., 1986, 15:931-934.

1064. Hodes, D. and Brunell, P.A. Mumps antibody: placental transfer and disappearance during the first year of life. Pediatrics, 1970, 45:99-101.

1065. Hoejenbos, E. and Stronk, M.G. In quest of toxoplasmosis as a cause of mental deficiency. Psychiatr. Neurol. Neurochir., 1966, 69:33-41.

1066. Hoekelman, R.A. and Anderson, V.M. Congenital thrombocytopenia, hepatosplenomegaly, and growth retardation. Am. J. Dis. Child, 1982, 136:258-264.

1067. Hogan, M.J., Kimura, S.J., Lewis, A. and Zweigart, P.A. Early and delayed ocular manifestations of congenital toxoplasmosis. Trans. Amer. Ophthalmol., 1957, 55:275-296.

1068. Hogg, G.G., Lenghaus, C. and Forman, A.J. Experimental porcine parvovirus infection of foetal pigs resulting in abortion, histological lesions and antibody formation. J. Comp. Pathol., 1977, 87:539-549.

1069. Holowach, J., Thurston, D.L. and Becker, B. Congenital defects in infants following mumps during pregnancy. A review of the literature and a report of chorioretinitis due to fetal infection. J. Pediat., 1957, 50:689.

1070. Holzbach, R.T. Australia antigen hepatitis in pregnancy. Evidence against transplacental transmission of Australia antigen in early and late pregnancy. Arch. Intern. Med., 1972, 130:234-236.

1071. Hoover, E.A. and Griesemer, R.A. Experimental feline herpesvirus infection in the pregnant cat. Am. J. Pathol., 1971, 65:173-188.

1072. Hoover, E.A. and Griesemer, R.A. Pathogenicity of feline viral rhinotracheitis virus and effect on germfree cats, growing bone, and the gravid uterus. J. Am. Vet. Med. Assoc., 1971, 158:929.

1073. Hoover, E.A., Schaller, J.P., Mathes, L.E. and Olsen, R.G. Passive immunity to feline leukemia: evaluation of immunity from dams naturally infected and experimentally vaccinated. Infect. Immun., 1977, 16:54-59.

1074. Hope Simpson, R.E. Rubella and congenital malformations. Lancet, 1944, 1:483.

1075. Hopkins, P. Rubella vaccination. Br. Med. J., 1979, 1:1712.

1076. Hore, D.E., Smith, H.V., Snowdon, W.A. and Uren, J. An investigation of viral and bacterial agents associated with ovine perinatal mortality. Aust. Vet. J., 1973, 49:190-195.

1077. Horn, P. Poliomyelitis in pregnancy. A twenty-year report from Los Angeles County, California. Obst. Gyn., 1955, 6:121.

1078. Horner, G.W. Serological relationship between abortifacient and respiratory strains of equine herpesvirus type 1 in New Zealand. N.Z. Vet. J., 1981, 29:7-8.

1079. Horner, G.W. and Hunter, R. Isolation of porcine parvovirus from aborted piglets. N.Z. Vet. J., 1977, 25:25-26.

1080. Horstmann, D.M. Rubella and the rubella syndrome: problems and progress. Med. Clin. North Am., 1967, 51:587-598.

1081. Horstmann, D.M. First-trimester exposure to rubella. J.A.M.A., 1968, 204:1153-1154.

1082. Horstmann, D.M. Discussion paper: the use of primates in experimental viral infections -- rubella and the rubella syndrome. Ann. N.Y. Acad. Sci., 1969, 162:594-597.

1083. Horstmann, D.M. Viral infections in pregnancy. Yale J. Biol. Med., 1969, 42:99-112.

1084. Horstmann, D.M. Rubella and the rubella syndrome: clinical epidemiology and prospects for control by vaccination. S. Afr. Med. J., 1970, 44:383-387.

1085. Horstmann, D.M. Rubella. Clin. Obstet. Gynecol., 1982, 25:585-597.

1086. Horstmann, D.M., Banatvala, J.E., Riordan, J.T., Payne, M.C., Whittemore, R., Opton, E.M. and Florey, C. DuVe. Maternal rubella and the rubella syndrome in infants. Epidemiologic, clinical, and virologic observations. Am. J. Dis. Child, 1965, 110:408-415.

1087. Horstmann, D.M., Liebhaber, H. and Kohorn, E.I. Post-partum vaccination of rubella -- susceptible women. Lancet, 1970, 2:1003-1006.

1088. Horstmann, D.M., Pajot, T.G. and Liebhaber, H. Epidemiology of rubella. Subclinical infection and occurrence of reinfection. Amer. J. Dis. Child, 1969, 118:133.

1089. Hosking, C.S., Ryman, C. and Wilkins, B. The nerve deaf child-intrauterine rubella or not? Arch. Dis. Child, 1983, 58:327-329.

1090. Hossain, A., Bakir, T.M. and Ramia, S. Immune status to congenital infections by TORCH agents in pregnant Saudi women. J. Trop. Pediatr., 1986, 32:83-86.

1091. Howard, C.J., Brownlie, J. and Thomas, L.H. Prevalence of bovine virus diarrhoea virus viraemia in cattle in the UK. Vet. Rec., 1986, 119:628-629.

1092. Howard, D.R. Transplacental transmission of rabies virus from a naturally infected skunk. Am. J. Vet. Res., 1981, 42:691-692.

1093. Hsiung, G.D., Fong, C.K. and Gross, P.A. Oncogenic potential of guinea pig herpes - and C-type viruses. Cancer Res., 1973, 33:1436-1442.

1094. Hsu, F.S., Chu, R.M., Lee, R.C. and Chu, S.H. Placental lesions caused by pseudorabies virus in pregnant sows. J. Am. Vet. Med. Assoc., 1980, 177:636-641.

1095. Hsu, F.S. and Lee, R.C. Use of hyperimmune serum, vaccination, and certain management procedures for control of pseudorabies in swine. J. Am. Vet. Med. Assoc., 1984, 184:1463-1466.

1096. Hsu, S.T., Chang, L.C., Lin, S.Y., Chuang, T.Y., Ma, C.H., Inoue, Y.K. and Okuno, T. The effect of vaccination with a live attenuated strain of Japanese encephalitis virus on stillbirths in swine in Taiwan. Bull. W.H.O., 1972, 46:465-471.

1097. Huang, J., Gentry, R.F. and Zarkower, A. Experimental infection of pregnant sows with porcine enteroviruses. Am. J. Vet. Res., 1980, 41:469-473.

1098. Huang, Y.S., Chernoff, N., Kavlock, R.J. and Kawanishi, C.Y. The effects of maternal murine cytomegalovirus infection on the mouse conceptus at different gestational stages. Teratogenesis Carcinog. Mutagen., 1986, 6:331-338.

1099. Hubbard, T.W. Varicella occurring in an infant twenty-four hours after birth. Br. Med. J., 1878, 1:822.

1100. Hubbert, W.T., Bryner, J.H., Estes, P.C. and Foley, J.W. Bovine bluetongue: viral isolation from placentome and serologic survey in pregnant cows. Am. J. Vet. Res., 1972, 33:1879-1882.

1101. Hubbert, W.T., Bryner, J.H., Fernelius, A.L., Frank, G.H. and Estes, P.C. Viral infection of the bovine fetus and its environment. Arch. Gesamte Virusforsch., 1973, 41:86-98.

1102. Hubbert, W.T., Fernelius, A.L., Maaten, M.H. Vander, Estes, P.C. and Bryner, J.H. Concurrent dual viral infection of a bovine foetus. Vet. Rec., 1973, 93:45-46.

1103. Hudson, A.W. and McFarland, C. Disseminated herpes simplex in a newborn. A consequence of infection in the mother. J.A.M.A., 1969, 208:859-861.

1104. Hudson, C.N. AIDS and obstetrics. Lancet, 1987, 1:388.

1105. Hug, G. Genetic factors and autoimmunity in viral hepatitis. Am. J. Clin. Pathol., 1976, 65:870-875.

1106. Hughes-Davies, T.H. Can E.B. virus affect the fetus? Lancet, 1970, 1:145.

1107. Hughes-Davies, T.H. Prevention of intrauterine rubella. Lancet, 1973, 1:943.

1108. Hughes, D.W., Parkinson, R., Beveridge, J., Reid, R.R. and Murphy, A.M. The expanded congenital rubella syndrome: report of two cases with neonatal purpura and review of the recent literature. Med. J. Aust., 1967, 1:420-425.

1109. Hughesdon, M.R. Congenital tuberculosis. Arch. Dis. Child, 1946, 21:121.

1110. Huikeshoven, F.J., Wallenburg, H.C. and Jahoda, M.G. Diagnosis of severe fetal cytomegalovirus infection from amniotic fluid in the third trimester of pregnancy. Am. J. Obstet. Gynecol., 1982, 142:1053-1054.

1111. Hultquist, G., Nordvall, S. and Sundstrom, C. Insulitis in cytomegalovirus infection in a newborn infant. Ups. J. Med. Sci., 1973, 78:139-144.

1112. Hunter, K., Stagno, S., Capps, E. and Smith, R.J. Prenatal screening of pregnant women for infections caused by cytomegalovirus, Epstein-Barr virus, herpesvirus, rubella and Toxoplasma gondii. Am. J. Obstet. Gynecol., 1983, 145:269-273.

1113. Hurley, R. Viral infections in pregnancy. Midwife Health Visit Community Nurse, 1977, 13:117-119.

1114. Hutchinson, D.L., Iskander, G.B., Gaffney, P.C. an Maxwell, N.G. Maternal and neonatal jaundice and hepatitis associated with intrauterine fetal transfusion. J.A.M.A., 1965, 194:199-201.

1115. Hutchinson, D.N. and Thompson, K.M. Continuous rabbit kidney cell cultures in the diagnosis of rubella infections. Mon. Bull. Minist. Health (Lond.), 1965, 24:385-391.

1116. Hutchison, D.S., Smith, R.E. and Haughton, P.B. Congenital herpetic keratitis. Arch. Ophthalmol., 1975, 93:70-73.

1117. Hutfield, D.C. Fatal effects of herpes genitalis on the newborn infant. J. Obstet. Gynaecol. Br. Commonw., 1966, 73:1020-1022.

1118. Hutto, C., Arvin, A., Jacobs, R., Steele, R., Stagno, S., Lyrene, R., Willett, L, Powell, D., Andersen, R., Werthammer, J. et al. Intrauterine herpes simplex virus infections. J. Pediatr., 1987, 110:97-101.

1119. Hwang, L.Y., Roggendorf, M., Beasley, R.P. and Deinhardt, F. Perinatal transmission of hepatitis B virus: role of maternal HBeAg and anti-HBc lgM. J. Med. Virol., 1985, 15:265-269.

1120. Hyatt, H. Relationship of maternal mumps to congenital defects and fetal deaths, and to maternal morbidity and mortality. Amer. Pract. Dig. Treat., 1961, 12:359.

# I

1121. Icart, J. and Didier, J. Infections due to Epstein-Barr virus during pregnancy. J. Infect. Dis., 1981, 143:499.

1122. Icart, J., Didier, J., Dalens, M., Chabanon, G. and Boucays, A. Prospective study of Epstein-Barr virus (EBV) infection during pregnancy. Biomedicine, 1981, 34:160-163.

1123. Ichida, F., Yoshikawa, A., Tokunaga, A. and Takeuchi, S. Hepatitis B immunoglobulin (HBIG) efficacy in the prevention of perinatal transmission of hepatitis B virus (HBV) infection. Gastroenterol. Jpn., 1983, 18:56-59.

1124. Ida, N., Ikawa, Y. and Oba, Y. Studies on hypodiploid virus-induced mouse leukemias and the vertical transmission of MSV-Moloney. Bibl. Haematol., 1970, 221-233.

1125. Ida, N., Moloney, J.B. and Taylor, G. Vertical transmission of Moloney leukemia virus in A-LN and BALB-c strains in mice. Gann., 1965, 56:239-249.

1126. Iida, T., Tajima, M. and Murata, Y. Transmission of maternal antibodies to Sendai virus in mice and its significance in enzootic infection. J. Gen. Virol., 1973, 18:247-254.

1127. Imamura, M., Phillips, P.E. and Mellors, R.C. The occurrence and frequency of type C virus-like particles in placentas from patients with systemic lupus erythematosus and from normal subjects. Am. J. Pathol., 1976, 83:383-394.

1128. Inaba, N., Ijichi, M., Ohkawa, R. and Takamizawa, H. Placental transmission of hepatitis B e antigen and clinical significance of hepatitis B e antigen titers in children born to hepatitis B antigen-positive carrier women. Am. J. Obstet. Gynecol., 1984, 149:580-581.

1129. Ingalls, T.H. Rubella vaccines and the immunity gap. Lancet, 1979, 1:831.

1130. Ingalls, T.H. and Horne, H.W., Jr. Immunization of women with rubella (RA27-3) vaccine administered intranasally. Lancet, 1971, 1:830-832.

1131. Inouye, S. and Kono, R. Hemagglutination-inhibiting (HI) antibody response against rubella virus in non-pregnant and pregnant rabbits and transmission of antibody to young. J. Immunol., 1972, 108:370-376.

1132. Ioachim, H.L. Prevention of Gross virus-induced leukemia in progeny of immunized female rats. Cancer Res., 1970, 30:2661-2664.

1133. Ioachim, H.L. Gross virus leukemia in the rat: kinetics of infection and immunity. Bibl. Haematol., 1973, 39:288-295.

1134. Ioachim, H.L. Perinatal viral carcinogenesis: the role of immunity. Natl. Cancer Inst. Monogr., 1979, 193-204.

1135. Ioachim, H.L., Gimovsky, M.L., Keller, S.E. and Gross, L. Maternal vaccination with formalin-inactivated Gross lymphoma virus in rats and transfer of immunity to offspring. Proc. Soc. Exp. Biol. Med., 1973, 144:376-379.

1136. Ioachim, H.L., Keller, S.E., Gimovsky, M.L. and Shepherd, S. Immunity to viral leukemia in rats and the routes of transmission from mother to offspring. Cancer Res., 1973, 33:547-550.

1137. Ip, H. and Corner, B.D. Thrombocytopenic purpura in cytomegalovirus mononucleosis. Lancet, 1973, 2:621.

1138. Isacsohn, M., Harlap, S., Nishmi, M. and Swartz, T.A. Rubella in Jerusalem. 1. Seroepidemiologic findings in children born after the 1972 rubella epidemic. Isr. J. Med. Sci., 1978, 14:954-961.

1139. Isacsohn, M., Rudensky, B., Eidelman, A.I., Abulafia, Y., Paz, E., Erlichman, M., Sayar, Y., Nubani, N. and Malachi, D. Cytomegalovirus infection and disease in the Jerusalem area. Isr. J. Med. Sci., 1983, 19:916-920.

1140. Ishiguro, T. and Ozaki, Y. Neutralizing antibodies against herpes simplex virus in maternal and umbilical cord serum. Acta. Obstet. Gynecol. Scand., 1979, 58:413-414.

1141. Ishii, S. and Ishitani, R. Equine infectious anemia. Adv. Vet. Sci. Comp. Med., 1975, 19:195-222.

1142. Ishimoto, A. and Ito, Y. Presence of antibody against mouse fetal antigen in the sera from C57BL-6 mice immunized with Rauscher leukemia. Cancer Res., 1972, 32:2332-2337.

1143 Ishimoto, A., Suzuki, Y., Yoshida, T. and Ito, Y. Further studies on mouse fetal antigen cross-reactive with Rauscher leukemia. Cancer Res., 1974, 34:2338-2343.

1144. Ishimoto, A., Suzuki, Y., Yoshida, T. and Ito, Y. Mouse fetal antigens in Rauscher leukemia. Bibl. Haematol., 1975, 311-317.

1145. Issel, C.J. Maternal antibody to Jamestown Canyon virus in white-tailed deer. Am. J. Trop. Med. Hyg., 1974, 23:242-245.

1146. Issel, C.J., Adams, W.V., Jr. and Foil, L.D. Prospective study of progeny of inapparent equine carriers of equine infectious anemia virus. Am. J. Vet. Res., 1985, 46:1114-1116.

1147. Iwai, H., Itoh, T. and Shumiya, S. Persistence of Sendai virus in a mouse breeder colony and possibility to re-establish the virus free colonies. Jikken-Dobutsu, 1977, 26:205-212.

1148. Iwakata, S., Rhodes, A.H. and Labzoffsky, N.A. Laboratory diagnosis of rubella virus infections. Can. Med. Assoc. J., 1973, 108:894-895.

1149. Iwarson, S., Norkrans, G., Hermodsson, S. and Nordenfelt, E. Passive -- active immunization in a neonate treated with repeated doses of high-titred hepatitis B immune globulin. Scand. J. Infect. Dis., 1979, 11:167-169.

# J

1150. Jack, I. Leukocyte viremia and intrauterine infection. Am. Heart J., 1970, 80:291-293.

1151. Jackson, A.D.M. and Fisch, L. Deafness following maternal rubella: results of a prospective investigation. Lancet, 1958, 2:1241-1244.

1152. Jackson, R. Herpes zoster in early pregnancy. Arch. Dermatol., 1975, 3:1658.

1153. Jacob, A.J., Epstein, J., Madden, D.L. and Sever, J.L. Genital herpes infection in pregnant women near term. Obstet. Gynecol., 1984, 63:480-484.

1154. Jacobs, J.W. and Peacock, D.B. Differentiation of actively and passively acquired complement-fixing antibodies in infants with respiratory syncytial virus infection. J. Med. Microbiol., 1970, 3:313-324.

1155. Jamrichova, O., Sokol, A. and Sevcik, A. Distribution of attenuated vaccine strains of pseudorabies virus in intraperitoneally infected swine foetuses. Acta. Virol. (Praha.), 1971, 15:286-292.

1156. Jankowski, M. and Gut, W. IgM antibodies in rubella infections: absorption with staphylococcal protein A of sera examined by immunofluorescence and haemagglutination inhibition. Acta. Microbiol. Pol., 1979, 28:293-299.

1157. Jankowski, M.G., Gut, W., Switalski, L., Imbs, D. and Kantoch, M. IgM fluorescence antibodies in sera of pregnant women exposed to rubella. Arch. Virol., 1979, 60:123-130.

1158. Janssen, D.L., Bartz, C.R., Bush, M., Marchwicki, R.H., Grate, S.J. and Montali, R.J. Parvovirus enteritis in vaccinated juvenile bush dogs. J. Am. Vet. Med. Assoc., 1982, 181:1225-1227.

1159. Jeffcott, L.B. and Rossdale, P.D. Practical aspects of equine virus abortion in the United Kingdom. Vet. Rec., 1976, 98:153-155.

1160. Jeffries, D.J. Management of herpes simplex infection during pregnancy and in the neonate. J. Hosp. Infect., 1983, 4:217-220.

1161. Jeffries, D.J. Cytomegalovirus infections in pregnancy. Br. J. Obstet. Gynaecol., 1984, 91:305-306.

1162. Jelinek, J.E. Localized accidental vaccination of the eye. Arch. Dermatol., 1969, 99:504.

1163. Jelliffe, D.B. Possible teratogenic effects of exotic viruses. Lancet, 1973, 1:774.

1164. Jelliffe, E.F.P. Low birth weight and malarial infection of the placenta. Bull. W.H.O., 1968, 38:69.

1165. Jenkins, N.A. and Copeland, N.G. High frequency germline acquisition of ecotropic MuLV proviruses in SWR/J-RF/J hybrid mice. Cell, 1985, 43:811-819.

1166. Jensen, S.C. and Eaton, S. The herpes simplex viruses: the facts health professionals need to know. Radiol. Technol., 1985, 56:302-305.

1167. Jenson, A.B., Groff, D.E., McConahey, P.J. and Dixon, F.J. Transmission of murine leukemia virus (Scripps) from parent to progeny mice as determined by p30 antigenemia. Cancer Res., 1976, 36:1228-1232.

1168. Jespersen, C.S., Lttauer, J. and Sagild, U. Measles as a cause of fetal defects. A retrospective study of ten measles epidemics in Greenland. Acta. Paediatr. Scand., 1977, 66:367-372.

1169. Jewett, J.F. Viral hepatitis. N. Engl. J. Med., 1973, 288:468-469.

1170. Jewett, J.F. Influenza pneumonia at term. N. Engl. J. Med., 1974, 291:256-257.

1171. Jewett, J.F. Committee on Maternal Welfare Herpes simplex encephalitis. N. Engl. J. Med., 1975, 292:531.

1172. Jhaveri, R.C., Verma, R.S., Rosenfeld, W., Salazar, D.J., Dosik, H. and Evans, H.E. Chromosomal abnormality in the newborns of hepatitis B surface antigen (HBsAg) carrier mothers. Clin. Pediatr. (Phila.), 1980, 19:66-68.

1173. Johnson, C. Cytomegalovirus infection. Midwife Health Visit Community Nurse, 1985, 21:166-168.

1174. Johnson, G.M. and Tudor, R.B. Diabetes mellitus and congenital rubella infection. Am. J. Dis. Child, 1970, 120:453-455.

1175. Johnson, K.P. Mouse cytomegalovirus: placental infection. J. Infect. Dis., 1969, 120:445-450.

1176. Johnson, K.P. Viral infections of the developing nervous system. Adv. Neurol., 1974, 6:53-67.

1177. Johnson, K.P. and Connor, W.S. Guinea pig cytomegalovirus: transplacental transmission. Brief report. Arch. Virol., 1979, 59:263-267.

1178. Johnson, R.E. Genital herpes and pregnancy. Am. Fam. Physician., 1986, 33:167-171.

1179. Johnson, R.H. Isolation of swine parvovirus in Queensland. Aust. Vet. J., 1973, 49:157-159.

1180. Johnson, R.H. and Collings, D.F. Experimental infection of piglets and pregnant gilts with a parvovirus. Vet. Rec., 1969, 85:446-447.

1181. Johnson, R.P. and Povey, R.C. Feline calicivirus infection in kittens borne by cats persistently infected with the virus. Res. Vet. Sci., 1984, 37:114-119.

1182. Johnson, R.T. Viral infections and malformations of the nervous system. Birth Defects, 1971, 7:56.

1183. Johnson, R.T. Effects of viral infection on the developing nervous system. N. Engl. J. Med., 1972, 287:599-604.

1184. Johnson, R.T. Hydrocephalus and viral infection. Dev. Med. Child Neurol., 1975, 17:807.

1185. Johnson, R.T. Viral infections and brain development. In: Berenberg SR, ed. Brain - fetal and infant. The Hague, Nijhoff Medical Division, 1977, 213-223.

1186. Johnson, R.T., Johnson, K.P. and Edmonds, C.J. Virus-induced hydrocephalus: development of aqueductal stenosis in hamsters after mumps infection. Science, 1967, 157:1066.

1187. Johnson, R.T., Shuey, H.E. and Buescher, E.L. Epidemic central nervous system disease of mixed enterovirus etiology. I. Clinical and epidemiologic description. Amer. J. Hyg., 1960, 71:321.

1188. Jones, C.T., Birch, C.R., Poller, L. and Dymock, I.W. Disseminated intravascular coagulation associated with hepatitis in late pregnancy. Gut., 1973, 14:891-894.

1189. Jones, D.E., dAvignon, M.B., Lawrence, R. and Latshaw, R.F. Burkitt's lymphoma: obstetric and gynecologic aspects. Obstet. Gynecol., 1980, 56:533-536.

1190. Jones, E. Cirrhosis of liver in infants and children. South Med. J., 1967, 60:57-62.

1191. Jones, J.F., Ray, C.G. and Fulginiti, V.A. Perinatal mumps infection. J. Pediatr., 1980, 96:912-914.

1192. Jones, M.J.U., Kolb, M., Votava, H.J., Johnson, R.L. and Smith, T.F. Case reports. Intrauterine ECHOvirus type II infection. Mayo. Clin. Proc., 1980, 55:509-512.

1193. Joo, H.S., Dee, S.A., Molitor, T.W. and Thacker, B.J. In utero infection of swine fetuses with infectious bovine rhinotracheitis virus (bovine herpesvirus-1). Am. J. Vet. Res., 1984, 45:1924-1927.

1194. Joo, H.S., Donaldson-Wood, C.R. and Johnson, R.H. Rapid diagnostic techniques for detection of porcine parvovirus infection in mummified foetuses. Aust. Vet. J., 1976, 52:51-52.

1195. Joo, H.S., Donaldson-Wood, C.R., Johnson, R.H. and Campbell, R.S. Pathogenesis of porcine parvovirus infection: pathology and immunofluorescence in the foetus. J. Comp. Pathol., 1977, 87:383-391.

1196. Joo, H.S. and Johnson, R.H. Observations on rapid diagnosis of porcine parvovirus in mummified foetuses. Aust. Vet. J., 1977, 53:106-107.

1197. Joos, T.H. Are congenital virus infections possible in successive pregnancies? Mich. Med., 1972, 71:33-34.

1198. Joosten, R. and Sturner, K.H. Hepatitis and pregnancy. Risks for the newborn. Immunoprophylaxis of vertically transmitted hepatitis. J. Perinat. Med., 1981, 9:115-123.

1199. Jordan, M., Poole, C.A. and Fogel, B.J. Extropy of the bladder associated with congenital rubella syndrome. J. Fla. Med. Assoc., 1968, 55:98-99.

1200. Jorgenson, R.J., Salinas, C.F. and Levin, L.S. Dental anomalies in the Morquio syndrome. Am. J. Dis. Child, 1976, 130:566-567.

1201. Josey, W.E., Nahmias, A.J. and Naib, Z.M. Genital herpes simplex in historical perspective. Bull. N.Y. Acad. Med., 1971, 52:935.

1202. Josey, W.E., Nahmias, A.J., Naib, Z.M., Utley, P.M., McKenzie, W.J. and Coleman, M.T. Genital herpes simplex infection in the female. Am. J. Obstet. Gynecol., 1966, 96:493-501.

1203. Jovaisas, E., Koch, M.A., Schafer, A., Stauber, M. and Lowenthal, D. LAV/HTLV-III in 20-week fetus. Lancet, 1985, 2:1129.

1204. Jovicic, B. and Stojanovic, M. Malignant hepatitis during pregnancy. On the occasion of healing six cases of hepatic coma. Riv. Ostet. Ginecol., 1968, 23:73-80.

1205. Judson, F.N., Shaw, B.S. and Vernon, T.M., Jr. Mandatory premarital rubella serologic testing in Colorado. A preliminary report. J.A.M.A., 1974, 229:1200-1202.

1206. Justines, G., Sucre, H. and Alvarez, O. Transplacental transmission of Venezuelan equine encephalitis virus in horses. Am. J. Trop. Med. Hyg., 1980, 29:653-656.

# K


1207. Kahrs, R.F. The relationship of bovine viral diarrheamucosal disease to abortion in cattle. J. Am. Vet. Med. Assoc., 1968, 153:1652-1655.

1208. Kahrs, R.F., Scott, F.W. and LaHunte, A. Bovine viral diarrhea-mucosal disease, abortion, and congenital cerebellar hypoplasia in a dairy herd. J. Am. Vet. Med. Assoc., 1970, 156:851-857.

1209. Kahrs, R.F., Scott, F.W. and DeLahunt A. Congenital cerebella hypoplasia and ocular defects in calves following bovine viral diarrhea-mucosal disease infection in pregnant cattle. J. Am. Vet. Med. Assoc., 1970, 156:1443-1450.

1210. Kahrs, R.F., Scott, F.W. and DeLahunta, A. Epidemiological observations on bovine viral diarrhea--mucosal disease virus-induced congenital cerebellar hypoplasia and ocular defects in calves. Teratology, 1970, 3:181-184.

1211. Kahrs, R.F. and Ward, G.M. Bovine virus diarrhea abortion. Proc. U.S. Anim. Health Assoc., 1967, 71:493-499.

1212. Kalter, S.S., Heberling, R.L., Panigel, M., Brack, M. and Felsburg, P.J. Fetal infection of the baboon (Papio cynocephalus) with vaccinia virus. Proc. Soc. Exp. Biol. Med., 1973, 143:1022-1024.

1213. Kalter, S.S., Heberling, R.L., Smith, G.C., Panigel, M., Kraemer, D.C., Helmke, R.J. and Hellman, A. Vertical transmission of C-type viruses: their presence in baboon follicular oocytes and tubal ova. J. Natl. Cancer Inst., 1975, 54:1173-1176.

1214. Kampa, D., Schmitz, H., Doerr, H.W., Luthardt, T., Hillemanns, H.G. and Wurtele, A. IgM antibodies to cytomegalovirus during pregnancy. Arch. Virol., 1977, 53:177-184.

1215. Kangro, H.O., Griffiths, P.D., Huber, T.J. and Heath, R.B. Specific IgM class antibody production following infection with cytomegalovirus. J. Med. Virol., 1982, 10:203-212.

1216. Kangro, H.O., Jackson, C. and Heath, R.B. Comparison of radioimmunoassay and the gel filtration technique for routine diagnosis of rubella during pregnancy. J. Hyg. (Lond.), 1981, 87:249-255.

1217. Kanra, G. Risk of rubella in pregnancy. Turk. J. Pediatr., 1982, 23:99-103.

1218. Kanra, G. and Firat, T. Isolation of rubella virus from lens material in cases of congenital cataracts. J. Pediatr. Ophthalmol. Strabismus., 1979, 16:31-32.

1219. Kanra, G., Gurses, N., Secmeer, G., Unal, M. and Gunalp, A. TORCH agents in neonates and in infants in Turkey. Turk. J. Pediatr., 1979, 21:4-6.

1220. Kapikian, A.Z., Kim, H.W., Wyatt, R.G., Cline, W.L, Parrott, R.H., Chanock, R.M., Arrobio, J.O., Brandt, C.D., Rodriguez, W.J., Kalica, A.R. and VanKirk, D.H. Recent advances in the aetiology of viral gastroenteritis. Ciba. Found. Symp., 1976, 273-309.

1221. Kaplan, M.H., Klein, S.W., McPhee, J. and Harper, R.G. Group B coxsackievirus infections in infants younger than three months of age: a serious childhood illness. Rev. Infect. Dis., 1983, 5:1019-1032.

1222. Kaplan, R.P. Skin changes and diseases associated with pregnancy. Obstet. Gynecol., 1979, 54:268.

1223. Karesh, J.W., Kapur, S. and MacDonald, M. Herpes simplex virus and congenital malformations. South Med. J., 1983, 76:1561-1563.

1224. Kari, B. and Gehrz, R. Susceptibility of fetal guinea pig brain cell cultures to replicating guinea pig cytomegalovirus infection is increased with increasing fetal age: infection of astrocytes. J. Virol., 1986, 58:960-962.

1225. Karkinen-Jaaskelainen, M. Spatial and temporal restriction of mumps virus induced lesions in the developing chick lens. Acta Path. Microbiol. Scand., 1973, 243:5-52.

1226. Karkinen-Jaaskelainen, Saxen, L. Maternal influenza, drug consumption, and congenital defects of the central nervous system. Am. J. Obstet. Gynecol., 1974, 118:815-818.

1227. Karmody, C.S. Subclinical maternal rubella and congenital deafness. N. Engl. J. Med., 1968, 278:809-814.

1228. Karmody, C.S. Asymptomatic maternal rubella and congenital deafness. Arch. Otolaryngol., 1969, 89:720-726.

1229. Karn, K., Julian, T.M. and Ogburn, P.L., Jr. Fetal heart block associated with congenital cytomegalovirus infection. A case report. J. Reprod. Med., 1984, 29:278-280.

1230. Katami, K., Taguchi, F., Nakayama, M., Goto, N. and Fujiwara, K. Vertical transmission of mouse hepatitis virus infection in mice. Jpn. J. Exp. Med., 1978, 48:481-490.

1231. Kato, K. Antigenic similarity between simian virus 40-induced surface and fetal antigens in hamster. J. Natl. Cancer Inst., 1977, 58:259-262.

1232. Kato, H., Yoshizawa, Y., Ueda, K. and Nagayama, T. Place of maternal rubella infection in the etiology of congenital heart disease. Jap. Circ. J., 1973, 37:47-51.

1233. Katz, M. Is there mumps embryopathy? An unanswered question. Clin. Pediatr. (Phila.), 1967, 6:321-322.

1234. Katz, R.G., White, L.R. and Sever, J.L. Maternal and congenital rubella. Clin. Pediatr. (Phila.), 1968, 7:323-330.

1235. Katz, S.L. Other viruses associated with infections of the fetus and newborn infant. In: Krugman, S. and Gershon, A.A., ed. Infections of the fetus and the newborn infant. New York, Liss, 1975, 55-61.

1236. Katz, V.L., Cefalo, R.C., McCune, B.K. and Moos, M.K. Elevated second trimester maternal serum alpha-fetoprotein and cytomegalovirus infection. Obstet. Gynecol., 1986, 68:580-581.

1237. Kaul, A. Management strategies for viral infections of the fetus and newborn. Indian J. Pediatr., 1981, 48:627-634.

1238. Kawakami, T.G., Sun, L. and McDowell, T.S. Natural transmission of gibbon leukemia virus. J.N.C.I., 1978, 61:1113-1115.

1239. Kay, H.E.M., Peppersorn, M.E., Porterfield, J.A., McCarthy, K. and Taylor-Robinson, C.H. Congenital rubella infection of a human embryo. Br. Med. J., 1964, 2:166-167.

1240. Kaye, B.M., Rosner, D.C. and Stein, I. Sr. Viral diseases in pregnancy and their effect upon the embryo and fetus. Amer. J. Obst. Gyn., 1953, 65:109.

1241. Keen, G.A. and Lambrick, M. False-positive reactions in rubella-specific IgM determinations. S. Afr. Med. J., 1981, 60:947.

1242. Keir, E.H. Results of rubella in pregnancy. II. Hearing defects. Med. J. Aust., 1965, 2:691-698.

1243. Kelemen, G. Rubella and deafness. Arch. Otolaryngol., 1966, 83:520-532.

1244. Kemen, M.J., Jr. and Coggins, L. Equine infectious anemia: transmission from infected mares to foals. J. Am. Vet. Med. Assoc., 1972, 161:496-499.

1245. Kemper, T.L., Lecours, A.R., Gates, M.J. and Yakovlev, P.I. Retardation of the myelo- and cytoarchitectonic maturation of the brain in the congenital rubella syndrome. Res. Publ. Assoc. Res. Nerv. Ment. Dis., 1973, 51:23-62.

1246. Kendig, E.L. Jr. Prognosis of infants born of tuberculous mothers. Pediatrics, 1960, 26:97.

1247. Kendrick, J.W. Bovine viral diarrhea-mucosal disease virus infection in pregnant cows. Am. J. Vet. Res., 1971, 32:433-444.

1248. Kendrick, J.W. and Osburn, B.I. Immunologic response to the bovine fetus to inactivated infectious bovine rhinotracheitis-infectious pustular vulvovaginitis virus. Am. J. Vet. Res., 1973, 34:1567-1571.

1249. Kendrick, J.W., Schneider, L. and Straub, O.C. Placental reaction to the infectious bovine rhinotracheitis-infectious pustular vulvovaginitis virus. Am. J. Vet. Res., 1971, 32:1045-1051.

1250. Kendrick, J.W. and Straub, O.C. Infectious bovine rhinotracheitis-infectious pustular vulvovaginitis virus infection in pregnant cows. Am. J. Vet. Res., 1967, 28:1269-1282.

1251. Kew, M.C., Marcus, R.G., Macnab, G.M. and Bersohn, I. Hepatitis B (surface) antigen in mothers and their infants. S. Afr. Med. J., 1975, 49:1471-1473.

1252. Keys, T.F., Sever, J.L., Hewitt, W.L. and Gitnick, G.L. Hepatitis-associated antigen in selected mothers and newborn infants. J. Pediatr., 1972, 80:650-653.

1253. Khatri, J.V., Kulkarni, K.V., Vaishnav, P.R. and Merchant, S.M. Vertical transmission of hepatitis-B. Indian Pediatr., 1980, 17:957-962.

1254. Khudr, G. and Benirschke, K. Placental lesion in viral hepatitis. Obstet. Gynecol., 1972, 40:381-384.

1255. Khudr, G. and Olding, L. Cyclopia. Am. J. Dis. Child, 1973, 125:120-122.

1256. Khuroo, M.s., Teli, M.R., Skidmore, S., Sofi, M.A. and Khuroo, M.I. Incidence and severity of viral hepatitis in pregnancy. Am. J. Med., 1981, 70:252-255.

1257. Kibrick, S. Herpes simplex virus in breast milk. Pediatrics, 1979, 64:390.

1258. Kibrick, S. Herpes simplex infection at term. What to do with mother, newborn, and nursery personnel. J.A.M.A., 1980, 243:157-160.

1259. Kibrick, S. and Benirschke, K. Acute aseptic myocarditis and meningo-encephalitis in the newborn child infected with Coxsackievirus group B, type 3. N. Engl. J. Med., 1956, 255:883.

1260. Kibrick, S. and Loria, R.M. Rubella and cytomegalovirus. Current concepts of congenital and acquired infection. Pediatr. Clin. North Am., 1974, 21:513-526.

1261. Kikuchi, I. Genital herpes simplex at term necessitating cesarean section. J. Dermatol. (Tokyo), 1986, 13:228-229.

1262. Kilbrick, S. and Benirschke, K. Severe generalized disease (encephalo-hepatomyocarditis) occurring in the newborn period and due to infection with Coxsackievirus, group B. Pediatrics, 1958, 22:857.

1263. Kilbride, H.W., Merenstein, G.B., Weisman, L.E. and Hill, J.M. Herpes simplex genitalis infections. Am. J. Obstet. Gynecol., 1980, 137:149-150.

1264. Kilham, L. and Margolis, G. Transplacental infection of rats and hamsters induced by oral and parenteral inoculations of H-I and rat viruses (RV). Teratology, 1969, 2:111-123.

1265. Kilham, L. and Margolis, G. Pathogenesis of intrauterine infections in rats due to reovirus type 3. I. Virologic studies. Lab. Invest., 1973, 28:597-604.

1266. Kilham, L. and Margolis, G. Pathogenesis of intrauterine infections in rats due to reovirus type 3. II. Pathologic and fluorescent antibody studies. Lab. Invest., 1973, 28:605.

1267. Kilham, L. and Margolis, G. Congenital infections due to reovirus type 3 in hamsters. Teratology, 1974, 9:51-63.

1268. Kilham, L. and Margolis, G. Intrauterine infections induced by mumps virus in hamsters. Lab. Invest., 1974, 31:34-41.

1269. Kilham, L. and Margolis, G. Induction of congenital hydrocephalus in hamsters with attenuated and natural strains of mumps virus. J. Infect. Dis., 1975, 132:462.

1270. Kilham, L. and Margolis, G. Problems of human concern arising from animal models of intrauterine and neonatal infections due to viruses: a review. I. Introduction and virologic studies. Prog. Med. Virol., 1975, 20:113-143.

1271. Kilham, L., Margolis, G. and Colby, D.E. Congenital infections of cats and ferrets by feline panleukopenia virus manifested by cerebellar hypoplasia. Lab. Invest., 1967, 17:465.

1272. Kilham, L., Margolis, G. and Colby, E.D. Cerebellar ataxia and its congenital transmission in cats by feline panleukopenia virus. J. Am. Vet. Med. Assoc., 1971, 158:888.

1273. Kilham, L., Murphy, H.W. and Overman, J.R. Performance of mumps neutralization tests in suckling mice. J. Immunol., 1953, 71:183.

1274. Kilham, L. and Overman, J.R. Natural pathogenicity of mumps virus for suckling hamsters on intra-cerebral injection. J. Immunol., 1953, 70:147.

1275. Kimball, A.C., Kean, B.H. and Fuchs, F. Congenital toxoplasmosis: a prospective study of 4,048 obstetric patients. J. Obstet. Gynecol., 1971, 111:211-218.

1276. King, A.A. and Harkness, J.W. Viral contamination of foetal bovine serum. Vet. Rec., 1975, 97:16.

1277. King, R.L. Herpes simplex encephalitis in pregnancy. Am. J. Obstet. Gynecol., 1979, 135:1114-1115.

1278. King-Lewis, P.A. and Gardner, S.D. Congenital cytomegalic inclusion disease following intrauterine transfusion. Br. Med. J., 1969, 2:603-605.

1279. Kipps, A. Rubella in the Republic. S. Afr. Med. J., 1977, 52:948.

1280. Kitamura, T., Tanaka, Y., Tagaya, I., Oba, S. and Kaneko, Y. Transplacental transmission of poxvirus neutralizing antibody. Jap. J. Med. Sci. Biol., 1973, 26:45-48.

1281. Klauber, G.T., Flynn, F.J., Jr. and Altman, B.D. Congenital varicella syndrome with genitourinary anomalies. Urology, 1976, 8:153-156.

1282. Kleiman, M.B. and Lamb, G.A. Gonococcal arthritis in a newborn infant. Pediatrics, 1973, 52:285-287.

1283. Klein, E.B., Byrne, T. and Cooper, L.Z. Neonatal rubella in a breast-fed infant after postpartum maternal infection. J. Pediatr., 1980, 97:774-775.

1284. Kleinman, H., Prince, J.T., Mathey, W.E., Rosenfield, A.B., Bearman, J.E. and Syverton, J.T. ECHO 9 virus infection and congenital abnormalities. A negative report. Pediatrics, 1962, 29:261.

1285. Klemola, E., Leinikki, P., Saksela, E., Essen, R. Von. Pregnancy during convalescence from cytomegalovirus mononucleosis. Scand. J. Infect. Dis., 1972, 4:255-258.

1286. Klingeborn, B. Equine abortion (herpes) virus: evaluation of markers in a field vaccination trial. Appl. Microbiol., 1973, 26:566-569.

1287. Klingeborn, B. and Dinter, Z. Equine abortion (herpes) virus: properties of the hemagglutinin in virus suspensions. Virology, 1973, 56:164-171.

1288. Klingeborn, B. and Pertoft, H. Equine abortion (herpes) virus: purification and concentration of enveloped and deenveloped virus and envelope material by density gradient centrifugation in colloidal silica. Virology, 1972, 48:618-623.

1289. Kluge, J.P. and Mare, C.J. Swine pseudorabies: abortion, clinical disease, and lesions in pregnant gilts infected with pseudorabies virus (Aujeszky's disease). Am. J. Vet. Res., 1974, 35:991-995.

1290. Kluge, T., Skyberg, D. and Bjoro, K. Fatal herpes simplex infection in the newborn. Caesarean section as a prophylactin measure. Acta. Obstet. Gynecol. Scand., 1967, 46:369-377.

1291. Kniazeff, A.J., Rimer, V. and Gaeta, L. Gamma-globulin foetal bovine sera: significance in virology. Nature, 1967, 214:805-806.

1292. Knott, P.D., Welply, G.A. and Anderson, M.J. Serologically proved intrauterine infection with parvovirus. Br. Med. J. (Clin. Res.), 1984, 289:1660.

1293. Knox, E.G. Epidemiology of prenatal infections: an extension of the congenital rubella model. Stat. Med., 1983, 2:1-12.

1294. Knox, E.G., Stewart, A. and Kneale, G. Childhood leukaemia and mother-foetus infection. Br. J. Cancer, 1980, 42:158-161.

1295. Knox, G.E. Influence of infection on fetal growth and development. J. Reprod. Med., 1978, 21:352-358.

1296. Knox, G.E. Cytomegalovirus: patient counseling. Semin. Perinatol., 1983, 7:43-46.

1297. Ko, T.M., Lin, K.H., Ho, M.M., Hwang, M.F., Hwang, K.C., Hsieh, F.J., Yaung, C.L. and Chen, D.S. Perinatal transmission of hepatitis B virus in the Taoyuan area. Taiwan I. Hsueh Hui Tsa Chih., 1986, 85:341-351.

1298. Kobbermann, T., Clark, L. and Griffin, W.T. Maternal death secondary to disseminated herpesvirus hominis. Am. J. Obstet. Gynecol., 1980, 137:742-743.

1299. Kobayashi, S. and Mukai, N. Retinoblastoma-like tumors induced by human adenovirus type 12 in rats. Cancer Res., 1974, 34:1646-1651.

1300. Koch, M.J. and Brody, J.A. Lack of teratogenic effects in a presumed parvovirus epidemic. N. Engl. J. Med., 1974, 290:1437.

1301. Koff, R.S. Hepatitis in pregnancy. N. Engl. J. Med., 1986, 314:1581-1582.

1302. Kogon, A., Kronmal, R. and Peterson, D.R. The relationship between infectious hepatitis and Down's syndrome. Am. J. Public Health Nations Health, 1968, 58:305-311.

1303. Kohl, S. and Loo, L.S. The relative role of transplacental and milk immune transfer in protection against lethal neonatal herpes simplex virus infection in mice. J. Infect. Dis., 1984, 149:38-42.

1304. Kohl, S., Shaban, S.S., Starr, S.E., Wood, P.A. and Nahmias, A.J. Human neonatal and maternal monocyte-macrophage and lymphocyte-mediated antibody-dependent cytotoxicity to cells infected with herpes simplex. J. Pediatr., 1978, 93:206-210.

1305. Komuro, A., Hayami, M., Fujii, H., Miyahara, S. and Hirayama, M. Vertical transmission of adult T-cell leukaemia virus. Lancet, 1983, 1:240.

1306. Kondela, A., Loretu, K. and Mella, P.N. Isolation of Rift Valley fever virus from cattle abortions in Tanzania. Trop. Anim. Health Prod., 1985, 17:185-186.

1307. Konno, S., Moriwaki, M. and Nakagawa, M. Akabane disease in cattle: congenital abnormalities caused by viral infection. Spontaneous disease. Vet. Pathol., 1982, 19:246-266.

1308. Konno, S. and Nakagawa, M. Akabane disease in cattle: congenital abnormalities caused by viral infection. Experimental disease. Vet. Pathol., 1982, 19:267-279.

1309. Kono, R., Hayakawa, Y., Hibi, M. and Ishii, K. Experimental vertical transmission of rubella virus in rabbits. Lancet, 1969, 1:343-347.

1310. Koplan, J.P., Goldstein, J. and Foster, S.O. Congenital vaccinia: some doubts. Pediatrics, 1972, 50:971-972.

1311. Koppe, J.G., Kloosterman, G.J., DeRoever-Bonnet, H., Eckert-Stroink, J.A., Loewer-Sieger, D.H. and DeBruijne, J.I. Toxoplasmosis and pregnancy, with a long-term follow-up of the children. Eur. J. Obstet. Gynecol. Reprod. Biol., 1974, 4:101-110.

1312. Koroly, M.V. Differential diagnosis of viral hepatitis and cholestasis of pregnancy: report of three cases. J. Am. Osteopath. Assoc., 1982, 82:273-277.

1313. Korones, S.B., Ainger, L.E., Monif, G.R., Roane, J., Sever, J.L. and Fuste, F. Congenital rubella syndrome: study of 22 infants. Myocardial damage and other new clinical aspects. Am. J. Dis. Child, 1965, 110:434-440.

1314. Korones, S.B., Ainger, L.E., Monif, G.R., Roane, J., Sever, J.L. and Fuste, F. Congenital rubella syndrome. New clinical aspects with recovery of virus from affected infants. J. Pediat., 1965, 67:166.

1315. Korones, S.B., Todaro, J., Roane, J.A. and Sever, J.L. Maternal virus infection after the first trimester of pregnancy and status of offspring to 4 years of age in a predominantly Negro population. J. Pediatr., 1970, 77:245-251.

1316. Korsantiya, B.M. and Smorodintsev, A.A. Transplacental transmission of endogenous interferon in pregnant mice inoculated with influenza or Newcastle disease viruses. Nature, 1971, 232:560-561.

1317. Kort, B.A., Cefalo, R.C. and Baker, V.V. Fatal influenza A pneumonia in pregnancy. Am. J. Perinatol., 1986, 3:179-182.

1318. Koskiniemi, M.L., Vaheri, A., Manninen, V. and Nikki, P. Ascending myelitis with high antibody titer to herpes simplex virus in the cerebrospinal fluid. J. Neurol., 1982, 227:187-191.

1319. Krajden, S., Bishai, F., Huang, S.N., Ronald, A., Harris, F., Kempston, J., Anderson, J. and Keston, S. Clinicopathological study of fulminant hepatitis: coinfection with hepatitis B virus and delta agent. Can. Med. Assoc. J., 1986, 135:1282-1285.

1320. Krakowka, S. Transplacentally acquired microbial and parasitic diseases of dogs. J. Am. Vet. Med. Assoc., 1977, 171:750-753.

1321. Krakowka, S., Confer, A. and Koestner, A. Evidence for transplacental transmission of canine distemper virus: two case reports. Am. J. Vet. Res., 1974, 35:1251-1253.

1322. Krakowka, S., Hoover, E.A., Koestner, A. and Ketring, K. Experimental and naturally occurring transplacental transmission of canine distemper virus. Am. J. Vet. Res., 1977, 38:919-922.

1323. Krakowka, S., Long, D. and Koestner, A. Influence of transplacentally acquired antibody on neonatal susceptibility to canine distemper virus in gnotobiotic dogs. J. Infect. Dis., 1978, 137:605-608.

1324. Kraus, B.S., Ames, M.D. and Clark, G.R. Effects of maternal rubella on dental crown development. Clin. Pediatr. (Phila.), 1969, 8:204-215.

1325. Kravitz, H. Acute and chronic German measles. III Med. J., 1965, 128:616-621.

1326. Krech, U. and Jung, M. Epidemiologie, Virologie and virologische Diagnose der Cytomegalie. Klin. Wschr., 1973, 51:529.

1327. Krech, U., Konjajev, Z. and Jung, M. Congenital cytomegalovirus infection in siblings from consecutive pregnancies. Helv. Paediatr. Acta., 1971, 26:355-362.

1328. Krech, U. and Sonnabend, W. Australia antigen in newborn babies. Lancet, 1970, 1:779.

1329. Krech, U., Sonnabend, W. and Jung, M. The age distribution of hepatitis-associated antigen (HAA) in Switzerland. Vox Sang., 1970, 19:391-393.

1330. Krech, U., Sonnabend, W., Kistler, G. and Mader, A. Australia antigen in cord blood. Vox Sang., 1973, 24:55-60.

1331. Kreider, J.W. and Breedis, C. The susceptibility of fetal rat skin in different immunologic environments to neoplastic induction with Shope papilloma virus. Cancer Res., 1969, 29:989-993.

1332. Kresky, B. and Nauheim, J.S. Rubella retinitis. Am. J. Dis. Child, 1967, 113:305-310.

1333. Kriel, R.L., Gates, G.A., Wulff, H., Powell, N., Poland, J.D. and Chin, T.D. Cytomegalovirus isolations associated with pregnancy wastage. Am. J. Obstet. Gynecol., 1970, 106:885-892.

1334. Krill, A.E. Retinopathy secondary to rubella. Int. Ophthalmol. Clin., 1972, 12:89-103.

1335. Kristensson, K., Olsson, Y. and Sourander, P. Virus encephalitis: pathogenesis in the immature brain. Dev. Med. Child Neurol., 1974, 16:382-394.

1336. Krous, H.F., Turbeville, D.F. and Altshuler, G.P. Campomelic syndrome -- possible role of intrauterine viral infection. Teratology, 1979, 19:9-14.

1337. Krugman, S. Rubella: clinical and epidemiological aspects. Arch. ges. Virusforsch., 1965, 16:477.

1338. Krugman, S. Panel discussion on future of rubella virus vaccines. Am. J. Dis. Child, 1969, 118:382-396.

1339. Krugman, S. Viral hepatitis: recent developments and prospects for prevention. J. Pediatr., 1975, 87:1067-1077.

1340. Krugman, S. Present status of measles and rubella immunization in the United States: a medical progress report. Review article. J. Pediatr., 1977, 90:1-12.

1341. Krugman, S. and Ward, R. The rubella problem. Clinical aspects, risk of fetal abnormality and methods of prevention. J. Pediatr., 1954, 44:489-498.

1342. Kudesia, G. and Carrington, D. Subclinical rubella in pregnancy. Lancet, 1985, 1:700.

1343. Kuhn, B.S. and Cohen, S.M. Care of the HBV positive mother and her infant. Health Care Women Int., 1986, 7:329-340.

1344. Kumar, M.L., Gold, E., Jacobs, I.B., Ernhart, C.B. and Nankervis, G.A. Primary cytomegalovirus infection in adolescent pregnancy. Pediatrics, 1984, 74:493-500.

1345. Kumar, M.L., Gold, E., Jacobs, I.B., Ernhart, C.B. and Nankervis, G.A. Primary cytomegalovirus infection in pregnancy. J. Pediatr., 1984, 104:669-673.

1346. Kumar, M.L. and Nankervis, G.A. Cytomegalovirus infections. South Med. J., 1979, 72:854-861.

1347. Kumar, M.L., Nankervis, G.A., Cooper, A.R. and Gold, E. Acquisition of cytomegalovirus infection following exhange transfusion: a prospective study. Transfusion, 1980, 20:327-331.

1348. Kumar, M.L., Nankervis, G.A., Cooper, A.R. and Gold, E. Postnatally acquired cytomegalovirus infections in infants of CMV-excreting mothers. J. Pediatr., 1984, 104:669-673.

1349. Kumar, M.L., Nankervis, G.A. and Gold, E. Inapparent congenital cytomegalovirus infection: A follow-up study. N. Engl. J. Med., 1973, 28:1370-1372.

1350. Kumar, M.L., Nankervis, G.A., Jacobs, I.B., Ernhart, C.B., Glasson, C.E., McMillan, P.M. and Gold, E. Congenital and postnatally acquired cytomegalovirus infections: long-term follow-up. J. Pediatr., 1984, 104:674-679.

1351. Kumar, M.L. and Prokay, S.L. Experimental primary cytomegalovirus infection in pregnancy: timing and fetal outcome. Am. J. Obstet. Gynecol., 1983, 145:56-60.

1352. Kurata, K., Kurata, T. and Aoyama, Y. Vertical transmission of type 1 and type 2 Herpes simplex virus infection in hamsters. Jpn. J. Exp. Med., 1976, 46:187-195.

1353. Kurata, T., Kurata, K. and Aoyama, Y. Vertical transmission of type 2 herpes simplex virus infection in immunosuppressed hamsters. Jpn. J. Exp. Med., 1978, 48:437-443.

1354. Kurogi, H., Inaba, Y., Goto, Y., Miura, Y. and Takahashi, H. Serologic evidence for etiologic role of Akabane virus in epizootic abortion-arthrogryposis -hydraencephaly in cattle in Japan, 1972-1974. Arch. Virol., 1975, 47:71-83.

1355. Kurogi, H., Inaba, Y., Takahashi, E., Sato, K. and Goto, Y. Experimental infection of pregnant goats with Akabane virus. Natl. Inst. Anim. Health Q (Tokyo), 1977, 17:1-9.

1356. Kurogi, H., Inaba, Y., Takahashi, E., Sato, K., Omori, T., Miura, Y., Goto, Y., Fujiwara, Y., Hatano, Y., Kodama, K., Fukuyama, S., Sasaki, N. and Matumoto, M. Epizootic congenital arthrogryposis-hydranencephaly syndrome in cattle: isolation of Akabane virus from affected fetuses. Arch. Virol., 1976, 51:67-74.

1357. Kurogi, H., Inaba, Y., Takahashi, E., Sato, K. and Satoda, K. Congenital abnormalities in newborn calves after inoculation of pregnant cows with Akabane virus. Infect. Immun., 1977, 17:338-343.

1358. Kurtz, J.B. and Sykes, D.W. Herpes-simplex virus (type 1) transmitted to a neonate. Lancet, 1974, 2:1580-1581.

1359. Kurtz, J.B., Tomlinson, A.H. and Pearson, J. Mumps virus isolated from a fetus. Br. Med. J. [Clin. Res.], 1982, 284:471.

1360. Kvasnicka, W.G. Reproductive problems associated with bluetongue virus activity in Nebraska. Prog. Clin. Biol. Res., 1985, 178:109-119.

# L


1361. Labzoffsky, N.A., Fish, N.A., Gyulai, E. and Roughley, F. A survey of toxoplasmosis among mentally retarded children. Can. Med. Assoc. J., 1965, 92:1026-1028.

1362. Lacorte, J.G. Persistence of influenza virus in mice inoculated by peritoneal route and their congenital transmission. Hospital (Rio de J), 1967, 71:141-144.

1363. Lacy, K. Fighting rubella. Midwife Health Visit Community Nurse, 1985, 21:172-174.

1364. Lafer, C.Z. and Morrison, A.N. Thrombocytopenic purpura progressing to transient hypoplastic anemia in a newborn with rubella syndrome. Pediatrics, 1966, 38:499-501.

1365. LaForet, E.G. and Lynch, C.L. Jr. Multiple congenital defects following maternal varicella. New Eng. J. Med., 1947, 236:534.

1366. Lagrew, D.C., Jr., Furlow, T.G., Hager, W.D. and Yarrish, R.L. Disseminated herpes simplex virus infection in pregnancy. Successful treatment with acyclovir. J.A.M.A., 1984, 252:2058-2059.

1367. Lam, K.M. and Hsiung, G.D. Further studies on the mode of transmission of herpes-like virus in guinea pigs. Proc. Soc. Exp. Biol. Med., 1971, 138:422-426.

1368. Lam, K.M. and Hsiung, G.D. Herpesvirus infection of guinea pigs. II. Transplacental transmission. Am. J. Epidemiol., 1971, 93:308-313.

1369. Lambert, G. and Fernelius, A.L. Bovine viral diarrhea virus and Escherichia coli in neonatal calf enteritis. Can. J. Comp. Med., 1968, 32:440-446.

1370. Lambert, H.P., Stern, H. and Wellsteed, A.J. Congenital rubella syndrome. Lancet, 1965, 2:826-827.

1371. Lamprecht, C. and Schauf, V. Difficulties in interpretation of rubella serology. J. Pediatr., 1983, 102:323-324.

1372. Lamprecht, C., Schauf, V., Warren, D., Nelson, K., Northrop, R. and Christiansen, M. An outbreak of congenital rubella in Chicago. J.A.M.A., 1982, 247:1129-1133.

1373. Lancaster, H.O. The epidemiology of deafness due to maternal rubella. Acta. Genet. (Basel), 1954, 5:12.

1374. Landsberger, E.J., Hager, W.D. and Grossman, J.H., III. Successful management of varicella pneumonia complicating pregnancy. A report of three cases. J. Reprod. Med., 1986, 31:311-314.

1375. Landsman, J.B., Grist, N.R. and Ross, C.A. ECHO 9 virus infection and congenital malformations. Brit. J. Prev. Soc. Med., 1964, 18:152.

1376. Lang, D. and Kummer, J.F. Demonstration of cytomegalovirus in semen. New Eng. J. Med., 1972, 287:756-758.

1377. Lang, D.J. The association of indirect inguinal hernia with congenital cytomegalic inclusion disease. Pediatrics, 1966, 38:913-915.

1378. Lang, D.J. Cytomegalovirus infections in pregnancy and the newborn. Clin. Obstet. Gynecol., 1970, 13:348-359.

1379. Lang, D.J. and Kummer, J.F. Demonstration of cytomegalovirus in semen. N. Engl. J. Med., 1972, 287:756.

1380. Lange, I., Rodeck, C.H., Morgan-Capner, P., Simmons, A. and Kangro, H.O. Prenatal serological diagnosis of intrauterine cytomegalovirus infection. Br. Med. J. [Clin. Res.], 1982, 284:1673-1674.

1381. Lansdown, A.B. Coxsackievirus B3 infection in pregnancy and its influence on foetal heart development. Br. J. Exp. Pathol., 1977, 58:378-385.

1382. Lansdown, A.B. Histological changes in the intestinal tract of pregnant mice infected with Coxsackievirus B3. Experientia., 1975, 31:327-328.

1383. Lansdown, A.B. Histological changes in the skeletal elements of developing rat foetuses following treatment with sodium salicylate. Fd. Cosmet. Toxicol., 1970, 8:647.

1384. Lansdown, A.B. Histological observations on epidermal development in fetal mice subject to intrauterine growth retardation due to maternal infections with Coxsackievirus B3. Biol. Neonate., 1975, 27:368-374.

1385. Lansdown, A.B. Histological observations on thymic development in fetal and newborn mammals subject to intra-uterine growth retardation. Biol. Neonat., 1977, 31:252.

1386. Lansdown, A.B. Influence of time of infection during pregnancy with coxsackievirus B3 on maternal pathology and foetal growth in mice. Br. J. Exp. Pathol., 1975, 56:119-123.

1387. Lansdown, A.B. Pathological changes in pregnant mice infected with Coxsackievirus B3 and given dietary casein hydrolysate supplement. Br. J. Exp. Pathol., 1975, 56:373-376.

1388. Lansdown, A.B. Influence of Coxsackievirus B infections on fetal growth in mice. Teratology, 1976, 13:291.

1389. Lansdown, A.B. Viral infections and diseases of the heart. Prog. Med. Virol., 1978, 24:70.

1390. Lansdown, A.B. and Brown, J.D. Immunization of mice against Coxsackievirus B3 and prevention of foetal growth retardation. Br. J. Exp. Pathol., 1976, 57:521-524.

1391. Lansdown, A.B. and Brown, J.D. Pathological observations on experimental cytomegalovirus infections in pregnancy. J. Pathol., 1978, 125:1-9.

1392. Lansdown, A.B. and Coid, C.R. Pathological changes in pregnant mice infected with Coxsackie B3 virus as a possible cause of retarded foetal development. Br. J. Exp. Pathol., 1974, 55:101-109.

1393. Lansdown, A.B., Coid, C.R. and Ramsden, D.B. Mitigation of virus-induced foetal growth retardation in mice by dietary casein hydrolysate. Nature, 1975, 254:599-600.

1394. Lansdown, A.B. and Ellaby, S.J. Histochemical demonstration of changes in liver cell enzymes in pregnant mice infected with Coxsackie B3 virus. Histochemistry, 1974, 40:175-180.

1395. Lansdown, A.B. and Grasso, P. Histological changes in the skeletal system of the developing quail embryo treated with sodium salicylate. Experientia, 1969, 25:885.

1396. Lapinleimu, K., Cantell, K., Koskimies, O. and Saxen, L. Association between maternal herpesvirus infections and congenital malformations. Lancet, 1974, 1:1127-1129.

1397. Lapinleimu, K., Koskimies, O., Cantell, K. and Saxen, L. Viral antibodies in mothers of defective children. Teratology, 1972, 5:345-351.

1398. Larke, R.P.B., Wheatley, E., Saroj, S. and Chernesky, M. Congenital cytomegalovirus infection in an urban Canadian community. J. Infect. Dis., 1980, 142:647-653.

1399. Larsen, A.E. and Porter, D.D. Pathogenesis of aleutian disease of mink: identification of nonpersistent infections. Infect. Immun., 1975, 11:92-94.

1400. Larsen, J.W., Jr. Influenza and pregnancy. Clin. Obstet. Gynecol., 1982, 25:599-603.

1401. Laurence, K.M., Carter, C.O. and David, P.A. Major central nervous systems malformations in South Wales. II pregnancy factors, seasonal variation, and social class effects. Br. J. Prev. Soc. Med., 1968, 22:212.

1402. Lawson, K.F. and Crawley, J.F. The ERA strain of rabies vaccine. Can. J. Comp. Med., 1972, 36:339-344.

1403. Lazar, R.M., Tamaroff, M., Nir, Y., Freund, B., O'Reilly, R., Kirkpatrick, D. and Kapoor, N. Language recovery following isolation for severe combined immunodeficiency disease. Nature, 1983, 306:54-55.

1404. Le, C.T., Chang, R.S. and Lipson, M.H. Epstein-Barr virus infections during pregnancy. A prospective study and review of the literature. Am. J. Dis. Child, 1983, 137:466-468.

1405. Lecatsas, G. Papillomavirus in urine in pregnancy and following renal transplantation. Prog. Clin. Biol. Res., 1983, 105:143-148.

1406. Leck, I. Further tests of the hypothesis that influenza in pregnancy causes malformations. H.S.M.H.A. Health Rep., 1971, 86:265-269.

1407. Leck, I., Hay, S., Wittie, J.J. and Greene, J.C. Malformations reported on birth certificates following influenza epidemics. Br. J. Prev. Soc. Med., 1970, 24:62.

1408. Leck, I., Hay, S., Wittie, J.J. and Greene, J.C. Malformations recorded on birth certificates following A2 influenza epidemics. Public Health Rep., 1969, 84:971-979.

1409. Leck, I. and Steward, J.K. Incidence of neoplasms in children born after influenza epidemics. Br. Med. J., 1972, 4:631-634.

1410. Lederman, M.M. Cell-mediated immunity and pregnancy. Chest., 1984, 86:6S-9S.

1411. Lee, A.K., Ip, H.M. and Wong, V.C. Mechanisms of maternal-fetal transmission of hepatitis B virus. J. Infect. Dis., 1978, 138:668-671.

1412. Lee, S.D., Lo, K.J., Wu, J.C., Tsai, Y.T., Wang, J.Y., Ting, L.P. and Tong, M.J. Prevention of maternal-infant hepatitis B virus transmission by immunization: the role of serum hepatitis B virus DNA. Hepatology, 1986, 6:369-373.

1413. Lehmann, J. Polymicrogyria and porencephaly after intrauterine cytomegaly virus infection. Neuropathol. Pol., 1982, 20:71-75.

1414. Lehmann, N.I., Ferris, A.A., Bennett, N.M. and Newman, J.W. Rubella: results of a serological survey of pregnant patients in Melbourne during 1968. Med. J. Aust., 1969, 1:1282-1283.

1415. Lehmann, N.I. and Gust, I.D. Viraemia in influenza. A report of two cases. Med. J. Aust., 1971, 2:1166.

1416. Leinikki, P., Granstrom, M.L., Santavuori, P. and Pettay, O. Epidemiology of cytomegalovirus infections during pregnancy and infancy. A prospective study. Scand. J. Infect. Dis., 1978, 10:165-171.

1417. Lejarraga, H. and Peckham, C.S. Birthweight and subsequent growth of children exposed to rubella infection in utero. Arch. Dis. Child, 1974, 49:50-54.

1418. Lenghaus, C., Forman, A.J. and Hale, C.J. Experimental infection of 35, 50 and 60 day old pig foetuses with porcine parvovirus. Aust. Vet. J., 1978, 54:418-422.

1419. Lennette, E.H., Schmidt, N.J. and Magoffin, R.L. Serology of rubella. Comparison of fluorescent antibody, complement fixation and neutralization tests for diagnosis of current infections and determination of sero-immunity. Calif. Med., 107:223-231.

1420. Lennette, E.H., Woodie, J.D. and Schmidt, N.J. A modified indirect immunofluorescent staining technique for the demonstration of rubella antibodies in human sera. J. Lab. Clin. Med., 1967, 69:689-695.

1421. Lerman, S.J., Lerman, L.M., Nankervis, G.A. and Gold, E. Accuracy of rubella history. Ann. Intern. Med., 1971, 74:97-98.

1422. Lerner, A.M. and Wilson, F.M. Virus cardiomyopathy. Prog. Med. Virol., 1973, 15:63.

1423. Levenson, I. Herpes simplex infection. Arch. Dermatol., 1984, 120:576.

1424. Leventon-Kriss, S., Rannon, L., Smetana, Z. and Yoffe, R. Increase in laboratory-confirmed cases of herpes genitalis and neonatal herpes infections in Israel. Isr. J. Med. Sci., 1983, 19:946-949.

1425. Levin, M.J., Oxman, M.N., Moore, M.G., Daniels, J.B. and Scheer, K. Diagnosis of congenital rubella in utero. N. Engl. J. Med., 1974, 290:1187-1188.

1426. Levine, J.B., Berkowitz, C.D. and St. Geme, J.W., Jr. Rubella virus reinfection during pregnancy leading to late-onset congenital rubella syndrome. J. Pediatr., 100:589-591.

1427. Levy, D.L. Persistent fetal tachycardia in utero prior to labor in an infant with congenital cytomegalic inclusion disease: case report. Am. J. Obstet. Gynecol., 1972, 112:859-860.

1428. Lewis, P.E., II, Cefalo, R.C. and Zaritsky, A.L. Fetal heart block caused by cytomegalovirus. Am. J. Obstet. Gynecol., 1980, 136:967-968.

1429. Lewkonia, I.K. and Jackson, A.A. Infantile herpes zoster after intrauterine exposure to varicella. Brit. Med. J., 1973, 3:149.

1430. Li, L. et al. The study of transmission of hepatitis B virus in utero. Chin. J. Prev. Med., 1986, 20:132-133.

1431. Lieberman, H.M. and Shafritz, D. Persistent hepatitis B virus infection and Hepatocellular carcinoma. Prog. Liver Dis., 1986, 8:395-415.

1432. Liebeschuetz, H.J. The effects of vaccination in pregnancy on the foetus. J. Obstet. Gynecol. Br. Cwth., 1964, 71:132.

1433. Liess, B., Orban, S., Frey, H.R., Trautwein, G., Wiefel, W. and Blindow, H. Studies on transplacental transmissibility of a bovine virus diarrhoea (BVD) vaccine virus in cattle. II. Inoculation of pregnant cows without detectable neutralizing antibodies to BVD virus 90 - 229 days before parturition (51st to 190th day of gestation). Zentralbl - Veterinarmed [B], 1984, 31:669-681.

1434. Lifschitz, M.H., Wilson, G.S., Smith, E.O. and Desmond, M.M. Factors affecting head growth and intellectual function in children of drug addicts. Pediatrics, 1985, 75:269-274.

1435. Lifson, A.R. and Rogers, M.F. Vertical transmission of human immunodeficiency virus. Lancet, 1986, 2:337.

1436. Liggins, G.C. and Phillips, L.I. Rubella embryopathy. An interim report on a New Zealand epidemic. Br. Med. J., 1963, 1:711-713.

1437. Liggitt, H.D., DeMartini, J.C. and Pearson, L.D. Immunologic responses of the bovine fetus to parvovirus infection. Am. J. Vet. Res., 1982, 43:1355-1359.

1438. Light, I.J. and Linnemann, C.C., Jr. Neonatal herpes simplex infection following delivery by cesarean section. Obstet. Gynecol., 1974, 44:496-499.

1439. Lima, M.T., Machado, R., Javech, C., Aguilera, A. and Cutie, E. Histopathological study of human placenta in women infected with rubella virus during pregnancy. Morphol.-Embryol. (Bucur), 1977, 23:257-260.

1440. Limbacter, J.P., Hill, M.E. and Janicki, P.C. Hypoplasia of the abdominal aorta associated with rubella syndrome. South Med. J., 1979, 72:617-619.

1441. Lin, H.H., Ohto, H., Etoh, T., Yoneyama, T., Kawana, T. and Mizuno, M. Studies on the risk factors of intrauterine infection of hepatitis B virus. Nippon Sanka Fujinka Gakkai Zasshi, 1985, 37:2393-2400.

1442. Lindquist, J.M., Plotkin, S.A., Shaw, L., Gilden, R.V. and Williams, M.L. Congenital rubella syndrome as a systemic infection. Studies of affected infants born in Philadelphia, U.S.A. Br. Med. J., 1965, 5475:1401-1406.

1443. Lindsay, J.S., Caruthers, D.G., Hemenway, W.G. and Harrison, M.S. Inner ear pathology following maternal rubella. Ann. Otol. Rhin. Laryng., 1953, 62:1201.

1444. Links, I.J., Whittington, R.J., Kennedy, D.J., Grewal, A. and Sharrock, A.J. An association between encephalomyocarditis virus infection and reproductive failure in pigs. Aust. Vet. J., 1986, 63:150-152.

1445. Linthicum, D.S., Horvath, L. and Carnegie, P.R. Radiometric assessment of experimental autoimmune encephalomyelitis in mice. Brain Res., 1979, 174:273-281.

1446. Linthicum, F.H., Jr. Viral causes of sensorineural hearing loss. Otolaryngol. Clin. North Am., 1978, 11:29-33.

1447. Lipps, B.V. and Mayor, H.D. Defective parvoviruses acquired via the transplacental route protect mice against lethal adenovirus infection. Infect. Immun., 1982, 37:200-204.

1448. Lipton, S.D., Bryant, J., Saed, F. and Fontillas, G. Fatal case of cytomegalovirus pneumonitis in a postpartum woman. Obstet. Gynecol., 1981, 57:670-673.

1449. Littlejohns, I.R. Encephalomyocarditis virus from stillborn pigs. Aust. Vet. J., 1984, 61:93.

1450. Lken, T. and Barlow, R.M. Border disease in Norway. Acta. Vet. Scand., 1981, 22:137-139.

1451. Lloyd-Evans, L.P. Rhinomune (rhinopneumonitis vaccine). Vet. Rec., 1982, 111:401.

1452. Lo, K.J., Tsai, Y.T., Lee, S.D., Wu, T.C., Wang, J.Y., Chen, G.H., Yeh, C.L., Chiang, B.N., Yeh, S.H. and Goudeau, A. et al. Immunoprophylaxis of infection with hepatitis B virus in infants born to hepatitis B surface antigen-positive carrier mothers. J. Infect. Dis., 1985, 152:817-822.

1453. Lockett, T. and Rudolph, J. Deaf-blind children with maternal rubella: implications for adult services. Am. Ann. Deaf., 1980, 125:1002-1006.

1454. Lomba, F., Bienfet, V. and Wellemans, G. IBR virus and occurrence of metritis at parturition in the bovine Belgian Blue White Breed. Br. Vet. J., 1976, 132:178-181.

1455. London, W.T. Epidemiologic lessons from viral hepatitis B. Ann. N.Y. Acad. Sci., 1984, 437:100-105.

1456. London, W.T., Curfman, B. and Sever, J.L. Mumps infection of the rhesus monkey fetus. Teratology, 1971, 4:234.

1457. London, W.T., Fuccillo, D.A., Anderson, B. and Sever, J.L. Concentration of rubella virus antigen in chondrocytes of congenitally infected rabbits. Nature, 1970, 226:172-173.

1458. London, W.T., Fuccillo, D.A., Sever, J.L. and Kent, S.G. Influenza virus as a teratogen in rhesus monkeys. Nature, 1975, 255:483-484.

1459. London, W.T., Levitt, N.H., Kent, S.G., Wong, V.G. and Sever, J.L. Congenital cerebral and ocular malformations induced in rhesus monkeys by Venezuelan equine encephalitis virus. Teratology, 1977, 16:285.

1460. London, W.T., Martinez, A.J., Houff, S.A., Wallen, W.C., Curfman, B.L., Traub, R.G. and Sever, J.L. Experimental congenital disease with simian cytomegalovirus in rhesus monkeys. Teratology, 1986, 33:323-331.

1461. London, W.T. and OConnell, A.P. Transplacental transmission of hepatitis B virus. Lancet, 1986, 1:1037-1038.

1462. Love, R.J. and Grewal, A.S. Reproductive failure in pigs caused by encephalomyocarditis virus. Aust. Vet. J., 1986, 63:128-129.

1463. Low, G.C. and Cooke, W.E. A congenital case of kala-azar. Lancet, 1926, 2:1209.

1464. Lowenthal, R.M., Funnell, C.F., Hope, D.M., Stewart, I.G. and Humphrey, D.C. Normal infant after combination chemotherapy including teniposide for Burkitt's lymphoma in pregnancy. Med. Pediatr. Oncol., 1982, 10:165-169.

1465. Lowthian, J.T. and Gillard, L.J., Jr. Varicella pneumonia in the gravid patient. J. La. State Med. Soc., 1980, 132:69-70.

1466. Luban, N.C., Kelleher, J.F. and Reaman, G.H. Altered distribution of T-lymphocyte subpopulations in children and adolescents with haemophilia. Lancet, 1983, 1:503-505.

1467. Lubiniecki, A.S., Cypress, R.H. and Hammon, W.M. Passive immunity for arbovirus infection. II. Quantitative aspects of naturally and artificially acquired protection in mice for Japanese (B) encephalitis virus. Am. J. Trop. Med. Hyg., 1973, 22:535-542.

1468. Lucas, M.H., Westcott, D.G., Edwards, S., Newman, R.H. and Swallow, C. Immunofluorescence and cell culture techniques in the diagnosis of viral infection of aborted bovine fetuses. Vet. Rec., 1986, 118:242-243.

1469. Ludwig, M.A. Pediatric management plan: rubella. Pediatr. Nurs., 1984, 10:295-296.

1470. Luedke, A.J. Effect of bluetongue virus on reproduction in sheep and cattle. Prog. Clin. Biol. Res., 1985, 178:71-78.

1471. Luedke, A.J., Jochim, M.M. and Jones, R.H. Bluetongue in cattle: effects of vector-transmitted bluetongue virus on calves previously infected in utero. Am. J. Vet. Res., 1977, 38:1697-1700.

1472. Luedke, A.J., Jochim, M.M. and Jones, R.H. Bluetongue in cattle: effects of Culicoides variipennis-transmitted bluetongue virus on pregnant heifers and their calves. Am. J. Vet. Res., 1977, 38:1687-1695.

1473. Luerti, M., Santini, A., Bernini, O., Castiglioni, M. and Ragni, M.C. ELISA antibodies to cytomegalovirus in pregnant patients: prevalence in and correlation with spontaneous abortion. Biol. Res. Pregnancy Perinatol., 1983, 4:181-183.

1474. Luisi, M. Smallpox vaccination and pregnancy. Am. J. Obstet. Gynecol., 1977, 128:700.

1475. Lukas, B., Wiesendanger, W. and Schmidt-Ruppin, K.H. Herpes genitatis in guinea pigs. I. Kinetic study in infection with herpesvirus hominis type 2. Arch. Virol., 1975, 49:1.

1476. Lundstrom, R. Rubella during pregnancy. A follow-up study of children born after an epidemic of rubella in Sweden, 1951. With additional investigations on prophylaxis and treatment of maternal rubella. Acta. Paediatr., 1962, 51:1-110.

1477. Lyday, J.H. Report of severe herpes zoster in a 13 1/2-year-old-boy whose chickenpox infection may have been acquired in utero. Pediatrics, 1972, 50:930.

1478. Lynch, F.W. Dermatologic conditions of the fetus, with particular reference to variola and vaccinia. Arch. Dermat. Syph. (Berl.), 1932, 26:997.

1479. Lyons, K.P. and Guze, L.B. Australia antigen associated hepatitis: radioimmunoassay in mother and infant. J.A.M.A., 1971, 215:981.

# M

1480. MacArthur, J.A., Mann, P.G., Oreffo, V. and Scott, G.B. Measles in monkeys: an epidemiological study. J. Hyg. (Lond.), 1979, 83:207-212.

1481. MacCallum, F.O. Viral hepatitis: some thoughts and a summary. Am. J. Med. Sci., 1975, 270:411-412.

1482. MacCarthy, D., Lindsay, M. and Morris, I. Children in hospital with mothers. Lancet, 1962, 1:603-608.

1483. Machado, R., Aguilera, A., Cutie, E., Hoyo, P. and Kouri, G. Morphologic alterations of experimental rubella syndrome in rabbits. Morphol. Embryol. (Bucur.), 1976, 22:41-45.

1484. MacKenzie, J.S. and Houghton, M. Influenza infections during pregnancy: association with congenital malformations and with subsequent neoplasms in children, and potential hazards of live virus vaccines. Bacteriol. Rev., 1974, 38:356-370.

1485. Mackenzie, J.S., Williams, K. and Papadimitriou, J. Influenza A virus and its influence on the outcome of pregnancy in the mouse. Dev. Biol. Stand., 1977, 39:489-496.

1486. Mackie, M.J., Fiscus, A.G. and Pallister, P. A study to determine causal relationships of toxoplasmosis to mental retardation. Amer. J. Epidemiol., 1971, 94:215-221.

1487. MacLachlan, N.J., Shore, C.E. and Osburn, B.I. Lymphocyte blastogenesis in bluetongue virus or Mycobacterium bovis-inoculated bovine fetuses. Vet. Immunol. Immunopathol., 1984, 7:11-18.

1488. MacMahon, B. and Yen, S. Influenza and neural-tube defects. Lancet, 1971, 2:260-261.

1489. Macrae, A.D. Rubella in pregnancy. Br. Med. J., 1974, 1:635.

1490. Madden, D.L., Fuccillo, D.A., Traub, R.G., Ley, A.C., Sever, J.L. and Beadle, E.L. Juvenile onset diabetes mellitus in pregnant women: failure to associate with Coxsackie B1-6, mumps, or respiratory syncytial virus infections. J. Pediatr., 1978, 92:959-960.

1491. Madden, D.L., Itis, J., Tzan, N. and Sever, J.L. Frequency of antibody to BK antigen in women whose children developed malignancies and women who developed detectable carcinoma in situ of the cervix during this pregnancy. Prog. Clin. Biol. Res., 1983, 105:149-156.

1492. Madden, D.L., Matthew, E.B., Dietzman, D.E., Purcell, R.H., Sever, J.L., Rostafinski, M. and Mata, A. Hepatitis and Down's syndrome. Am. J. Ment. Defic., 1976, 80:401-406.

1493. Maeck, J.V. and Phillips, C.A. Rubella vaccine program: its implications in obstetric practice. Am. J. Obstet. Gynecol., 1972, 112:513-518.

1494. Mahaffey, L.W. Histological lesions and virus isolation in rhinopneumonitis infection. Proc. R. Soc. Med., 1966, 59:1081-1083.

1495. Maheswari, T.R., Devi, R.S., Shenoy, K.T., Sathiavathy, N. and Balakrishnan, V. Markers of viral hepatitis in pregnancy. Indian J. Gastroenterol., 1985, 4:77-79.

1496. Maheswari, T.R., Shenoy, K.T., Devi, R.S., Sathiavathy, N. and Balakrishnan, V. A study of hepatitis B antigenemia in pregnant females and the risk of transmission to babies. Indian J. Gastroenterol., 1985, 4:83-84.

1497. Maidman, J.E. and Paton, J.B. Rubella in pregnancy. Obstet. Gynecol. Annu., 1974, 3:345-375.

1498. Mair, H.J. and Buchan, A.R. Rubella vaccination and termination of pregnancy. Br. Med. J., 1972, 4:271-273.

1499. Malfetano, J.H., Marin, A.C. and Malfetano, J.H., Jr. Laser treatment of condylomata acuminata in pregnancy. J. Reprod. Med., 1981, 26:574-576.

1500. Malkinson, M.H. Transference of maternal passive immunity to vaccinia virus in mice. Nature, 1967, 216:1117-1118.

1501. Mallia, C.P. and Nancekivell, A.F. Fulminant virus hepatitis in late pregnancy. Ann. Trop. Med. Parasitol., 1982, 76:143-146.

1502. Mandal, D. and Chowdhury, N.N. Viral hepatitis in pregnancy. J. Indian Med. Assoc., 1982, 79:96-98.

1503. Mandel, T.E., Jack, I. and Tait, B.D. HLA-DR typing of fetal human spleen and liver lymphoblastoid cells transformed by Epstein-Barr virus. Transplantation, 1982, 34:50-53.

1504. Mandelbaum, B. and Brough, A.J. Hepatitis following multiple intrauterine transfusions. Report of a case. Obstet. Gynecol., 1967, 30:188-191.

1505. Mann, J.M., Preblud, S.R., Hoffman, R.E., Brandling-Bennet, A.D., Hinman, A.R. and Herrmann, K.L. Assessing risks of rubella infection during pregnancy. A standardized approach. J.A.M.A., 1981, 245:1647-1652.

1506. Manuelidis, E.E. and Manuelidis, L. Experiments on maternal transmission of Creutzfeldt-Jakob disease in guinea pigs. Proc. Soc. Exp. Biol. Med., 1979, 160:233-236.

1507. Manzie, P.P. Herpes zoster. The GP view. Aust. Fam. Physician, 1982, 11:167-172.

1508. Marchese, C.A., Carbonara, A.O., Viora, E., LaProva, A. and Campogrande, M. Biopsy of chorionic villi for prenatal diagnosis. Acta. Obstet. Gynecol. Scand., 1984, 63:737.

1509. Marchette, N.J., Halstead, S.B., ORourke, T., Scott, R.M., Bancroft, W.H. and Vanopruks, V. Effect of immune status on Dengue 2 virus replication in cultured leukocytes from infants and children. Infect. Immun., 1979, 24:47-50.

1510. Margolis, G. and Kilham, L. Virus-induced cerebellar hypoplasia. Res. Publ. Assoc. Res. Nerv. Ment. Dis., 1968, 44:113-146.

1511. Margolis, G. and Kilham, L. Parvovirus infections, vascular endothelium, and hemorrhagic encephalopathy. Lab Invest., 1970, 22:478-488.

1512. Margolis, G. and Kilham, L. Pathogenesis of intrauterine infections in rats due to reovirus type 3. II. Pathologic and fluorescent antibody studies. Lab Invest., 1973, 28:605-613.

1513. Margolis, G. and Kilham, L. Problems of human concern arising from animal models of intrauterine and neonatal infections due to viruses: a review. II. Pathologic studies. Prog. Med. Virol., 1975, 20:144-179.

1514. Margolis, G. and Kilham, L. Induction of congenital hydrocephalus in hamsters with parainfluenza type 2 virus. Exp. Mol. Pathol., 1977, 27:235-248.

1515. Marion, R.W., Wiznia, A.A., Hutcheon, G. and Rubinstein, A. Human T-cell lymphotropic virus type III (HTLV-III) embryopathy. A new dysmorphic syndrome associated with intrauterine HTLV-III infection. Am. J. Dis. Child, 1986, 140:638-640.

1516. Marques-Dias, M.J., Harmant-VanRijckevorsel, G., Landrieu, P. and Lyon, G. Prenatal cytomegalovirus disease and cerebral microgyria: evidence for perfusion failure, not disturbance of histogenesis, as the major cause of fetal cytomegalovirus encephalopathy. Neuropediatrics, 1984, 15:18-24.

1517. Marquis, J.R. and Lee, J.K. Extensive central nervous system calcification in a stillborn male infant due to cytomegalovirus infection. Am. J. Roentgenol., 1976, 127:665-667.

1518. Marshall, W.C. Rubella. 2. Nurs. Times, 1967, 63:825-826.

1519. Marshall, W.C. The clinical impact of intrauterine rubella. Ciba. Found. Symp., 1972, 10:3-22.

1520. Marshall, W.C. Effects of rubella virus on the human fetus. Proc. R. Soc. Med., 1975, 68:369-371.

1521. Marshall, W.C. The development of immune mechanisms in the fetus and newborn. Major Probl. Clin. Pediatr., 1978, 17:259-277.

1522. Marshall, W.C. and Dudgeon, J.A. Australia antigen in a child with congenital malformations and in his mother. Am. J. Dis. Child, 1972, 123:378-379.

1523. Marshall, W.C. and Peckham, C.S. The management of herpes simplex in pregnant women and neonates. J. Infect., 1983, 6:23-29.

1524. Marshall, W.C., Peckham, C.S., Dudgeon, J.A., Sheppard, S., Smithells, R.W. and Weatherall, J.A. Parity of women contracting rubella in pregnancy. Implications with respect to rubella vaccination. Lancet, 1976, 1:1231-1233.

1525. Marshall, W.C, Sheppard, S., Stark, O., Milton, A. and Smithells, R.W. Rash in early pregnancy. Br. Med. J. [Clin. Res.], 1983, 287:609.

1526. Martell, M.A., Montes, F.C. and Alcocer, R. Transplacental transmission of bovine rabies after natural infection. J. Infect. Dis., 1973, 127:291-293.

1527. Martin, B.R. and Larson, K.A. Immune response of equine fetus to coliphage T2. Am. J. Vet. Res., 1973, 34:1363-1364.

1528. Marymond, J.H., Jr. and Herrmann, K.L. Rubella in pregnancy: review of current problems. Postgrad. Med., 1974, 56:167-172.

1529. Masterson, J. Respiratory complications of epidemic influenza. J. Ir. Med. Assoc., 1969, 62:37-40.

1530. Masur, H., Michelis, M.A., Greene, J.B., Onorato, I., VanDeStouwe, R.A., Holzman, R.S., Wormser, G., Brettman, L., Lange, M., Murray, H.W. and Cunningham-Rundles, S. An outbreak of community-acquired Pneumocystis carinii pneumonia: initial manifestations of a cellular immune dysfunction. New England Journal of Medicine, 1981, 305:1431-1438.

1531. Masurel, N., DeBruijne, J.I., Beuningh, H.A. and Schouten, H.J. Haemagglutination-inhibition antibodies against influenza A and influenza B in maternal and neonatal sera. J. Hyg. (Lond.), 1978, 80:13-19.

1532. Mata, L., Urrutia, J.J., Serrato, G., Mohs, E. and Chin, T.D. Viral infections during pregnancy and in early life. Am. J. Clin. Nutr., 1977, 30:1834-1842.

1533. Mathot, C. and Scher, S. Transmission through milk of antibodies against the Friend virus in mice. Nature, 1968, 219:82-83.

1534. Mathur, A., Arora, K.L. and Chaturvedi, U.C. Transplacental Japanese encephalitis virus (JEV) infection in mice during consecutive pregnancies. J. Gen. Virol., 1982, 59:213-217.

1535. Mathur, A., Arora, K.L., Rawat, S. and Chaturvedi, U.C. Japanese encephalitis virus latency following congenital infection in mice. J. Gen. Virol., 1986, 67:945-947.

1536. Mathur, A., Arora, K.L., Rawat, S. and Chaturvedi, U.C. Persistence, latency and reactivation of Japanese encephalitis virus infection in mice. J. Gen. Virol., 1986, 67:381-385.

1537. Mathur, A., Tandon, H.O., Mathur, K.R., Sarkari, N.B., Singh, U.K. and Chaturvedi, U.C. Japanese encephalitis virus infection during pregnancy. Indian J. Med. Res., 1985, 81:9-12.

1538. Mathur, A., Tripathi, R., Chaturvedi, U.C. and Mehra, P. Congenital rubella following inapparent rubella infection. Indian J. Med. Res., 1982, 75:469-473.

1539. Mathur, U., Bentley, D.W. and Hall, C.B. Concurrent respiratory synctytial virus and influenza. Ann. Intern. Med., 1980, 93:49-52.

1540. Matkin, N.D., Petersen, W.H. and Peck, J.E. Maternal rubella and hearing loss: a preliminary report on findings from the 1964 epidemic. Conn. Med., 1967, 31:850-854.

1541. Matsoukas, J.A. Viral antibody titers to rubella in coxa plana or Perthes' disease. Perthes diseases: is it the late osseous redidua of a minor prenatal rubella? Acta. Orthop. Scand., 1975, 46:957-962.

1542. Mayer, A.R. and Duff, P. Longitudinal screening for herpes simplex virus in obstetric patients. Milit. Med., 1982, 147:492-495.

1543. Mayer, T.R. Optimum use of rubella immunization. Am. Fam. Physician, 1981, 23:143-145.

1544. Mayor, H.D., Houlditch, G.S. and Mumford, D.M. Influence of adeno-associated satellite virus on adenovirus-induced tumours in hamsters. Nature [New Biol.], 1973, 241:44-46.

1545. Mazel, J.A., Schalm, S.W., DeGast, B.C., Nuijten, A.S., Heijtink, R.A., Botman, M.J., Banffer, J.R., Gerards, L.J., Zwijnenberg, J., Mettau, J. et al. Passive-active immunisation of neonates of HBsAg positive carrier mothers: preliminary observations. Br. Med. J. [Clin. Res.], 1984, 288:513-515.

1546. Mazzur, S. Behavior and disease: a possible approach. Am. J. Phys. Anthropol., 1970, 32:309-319.

1547. Mazzur, S. Menstrual blood as a vehicle of Australia-antigen transmission. Lancet, 1973, 1:749-751.

1548. McCabe, M.J., Pribram, H.F. and OLoughlin, B.J. Cytomegalic inclusion diseases; a clinico-pathological correlation. Neuroradiology, 1972, 3:177-181.

1549. McCallin, P.F., Fuccillo, D.A., Ley, A.C., Gilkeson, M.R., Traub, R.G. and Sever, J.L. Gammaglobulin as prophylaxis against rubella-induced congenital anomalies. Obstet. Gynecol., 1972, 39:185-189.

1550. McCance, D.J. and Mims, C.A. Transplacental transmission of polyoma virus in mice. Infect. Immun., 1977, 18:196-202.

1551. McCance, D.J. and Mims, C.A. Reactivation of polyoma virus in kidneys of persistently infected mice during pregnancy. Infect. Immun., 1979, 25:998-1002.

1552. McCarthy, K. Progress in prophylaxis: rubella. Ciba. Found. Symp., 1972, 10:165-177.

1553. McClurkin, A.W., Coria, M.F. and Smith, R.L. Evaluation of acetylethyleneimine-killed bovine viral diarrhea-mucosal disease virus (BVD) vaccine for prevention of BVD infection of the fetus. Proc. Annu. Meet. US. Anim. Health Assoc., 1975, 114-123.

1554. McCollum, R.W. and Zuckerman, A.J. Viral hepatitis: report on WHO informal consultation. J. Med. Virol., 1981, 8:1-29.

1555. McColm, B. Herpes virus hominus: case history of an infected baby. Aust. Nurses J., 1979, 9:38-39.

1556. McCracken, G.H., Ginsburg, C.M., Zweighaft, T.C. and Clahsen, J. Pharmacokinetics of rifampin in infants and children: relevance to prophylaxis against Haemophilus influenza type b disease. Pediatrics, 1980, 66:17-21.

1557. McCracken, G.H., Shinefield, H.R., Cobb, K. Congenital cytomegalic inclusion disease: A longitudinal study of 20 patients. Am. J. Dis. Child, 1969, 117:522-539.

1558. McCubbin, J.H. and Smith, J.S. How to diagnose rubella during pregnancy. Am. Fam. Physician, 1981, 23:205-208.

1559. McDonald, J.C. and Peckham, C.S. Gammaglobulin in prevention of rubella and congenital defect: a study of 30,000 pregnancies. Br. Med. J., 1967, 3:633-637.

1560. McDonald, P.J. Hepatitis B and pregnancy. Australas-Nurses J., 1981, 10:11-12.

1561. McDonald, R. Congenital syphilis has many faces. Clin. Pediat., 1970, 9:110.

1562. McDonald, R. and Heese, H. deV. The congenital rubella syndrome. S. Afr. Med. J., 1966, 40:470-472.

1563. McElwain, W.P. Serologic tests for rubella. J. Ky. Med. Assoc., 1973, 71:564-565.

1564. McFeely, R.A., Merritt, A.M. and Steraly, E.L. Abortion in a dairy herd vaccinated for infectious bovine rhinotracheitis. J. Am. Vet. Med. Assoc., 1968, 153:657-661.

1565. McGregor, J..A., Burns, J.C., Levin, M.J., Burlington, B. and Meiklejohn, G. Transplacental passage of influenza A/Bangkok (H3N2) mimicking amniotic fluid infection syndrome. Am. J. Obstet. Gynecol., 1984, 149:856-859.

1566. McGuire, T.C., Crawford, T.B. and Henson, J.B. Prevalence of antibodies to herpesvirus types 1 and 2, arteritis and infectious anemia viral antigens in equine serum. Am. J. Vet. Res., 1974, 35:181-185.

1567. McKendry, J.B. and Bailey, J.D. Congenital varicella associated with multiple defects. Can. Med. Assoc. J., 1973, 108:66-68.

1568. McKenzie, J.S. and Houghton, M. Influenza virus infections during pregnancy: association with congenital malformation and with subsequent neoplasms in children, and potential hazards of live virus vaccines. Bacteriol. Rev., 1974, 38:356.

1569. McKercher, D.G. Relationship of viruses to reproductive problems. J. Am. Vet. Med. Assoc., 1969, 154:1184-1191.

1570. McKercher, D.G., Wada, E.M. Ault, S.K. and Theis, J.H. Viral agents recovered from Ornithodoros coriaceus, a vector of epizootic bovine abortion. Am. J. Vet. Res., 1980, 41:803-805.

1571. McKinlay, W.J. Herpes zoster in pregnancy. Br. Med. J., 1980, 280:561-562.

1572. McKinnon, G.T. and Pringle, R.C. A serological study of antibody to Epstein-Barr virus in an Australian population. Med. J. Aust., 1974, 2:243-246.

1573. McLean, D.M., McNaughton, G.A., Givan, K.F., Best, J.M., Smith, P.A. and Coleman, M.A. Rubella virus infections during pregnancy, Toronto, 1963-1966. Can. Med. Assoc. J., 1966, 95:1174-1178.

1574. McSwiggan, D.A. and Taylor, C.E. Rubella antibody tests in pregnancy. Br. Med. J., 1976, 2:174.

1575. Meadow, S.R. Infectious hepatitis and stillbirth. Br. Med. J., 1968, 1:426.

1576. Medearis, D.N. Mouse cytomegalovirus infection. III. Attempts to produce intrauterine infections. Am. J. Hyg., 1964, 80:113.

1577. Medearis, D.N. Observations concerning human cytomegalovirus infection and disease. Bull Johns Hopkins Hosp., 1964, 114:181-211.

1578. Medearis, D.N. Jr. Human cytomegalovirus immunization prospects. N. Engl. J. Med., 1977, 296:1289-1290.

1579. Medearis, D.N. Jr. CMV immunity: imperfect but protective. N. Engl. J. Med., 1982, 306:985-986.

1580. Medearis, D.N. Jr. and Prokay, S.L. Effect of immunization of mothers on cytomegalovirus infection in suckling mice. Proc. Soc. Exp. Biol. Med., 1978, 157:523-527.

1581. Megafu, U. Jaundice in pregnancy aetiology, management and mortality at Enugu, Nigeria. East Afr. Med. J., 1981, 58:501-509.

1582. Megale, P., Angulo, J.J. and Pederneiras, C.A. Variola minor in Braganca Paulista county, 1956. Observations on the clinical course of variola minor and on pregnancy in women with the disease. Bull Soc. Pathol. Exot. Filiales, 1979, 72:11-20.

1583. Meguro, H., Kervina, M. and Wright, P.F. Antibody-dependent cell-mediated cytotoxicity against cells infected with respiratory syncytial virus: characterization of in vitro and in vivo properties. J. Immunol., 1979, 122:2521-2526.

1584. Mehrotra, R. Histopathological and immunohistochemical changes in placenta due to acute viral hepatitis during pregnancy. Indian J. Med. Res., 1986, 83:282-292.

1585. Melish, E.M. and Hanshaw, J.B. Congenital cytomegalovirus infections: Developmental progress of infants detected by routine screening. Am. J. Dis. Child, 1973, 126:190-194.

1586. Melnick, J.L. The general nature of viruses. Clin. Obstet. Gynecol., 1972, 15:858-876.

1587. Mendelow, D.A. and Lewis, G.C. Jr. Varicella pneumonia during pregnancy. Obstet. Gynecol., 1969, 33:98-99.

1588. Mengeling, W.L. Porcine parvovirus: frequency of naturally occurring transplacental infection and viral contamination of fetal porcine kidney cell cultures. Am. J. Vet. Res., 1975, 36:41-44.

1589. Mengeling, W.L. Prevalence of porcine parvovirus-induced reproductive failure: an abattoir study. J. Am. Vet. Med. Assoc., 1978, 172:1291-1294.

1590. Mengeling, W.L. Prenatal infection following maternal exposure to porcine parvovirus on either the seventh or fourteenth day of gestation. Can. J. Comp. Med., 1979, 43:106-109.

1591. Mengeling, W.L., Brown, T.T., Paul, P.S. and Gutekunst, D.E. Efficacy of an inactivated virus vaccine for prevention of porcine parvovirus-induced reproductive failure. Am. J. Vet. Res., 1979, 40:204-207.

1592. Mengeling, W.L. and Cutlip, R.C. Pathogenesis of in utero infection: experimental infection of five-week-old porcine fetuses with porcine parvovirus. Am. J. Vet. Res., 1975, 36:1173-1177.

1593. Mengeling, W.L. and Cutlip, R.C. Reproductive disease experimentally induced by exposing pregnant gilts to parvovirus. Am. J. Vet. Res., 1976, 37:1393-1400.

1594. Mengeling, W.L., Cutlip, R.C., Wilson, R.A., Parks, J.B. and Marshall, R.F. Fetal mummification associated with porcine parvovirus infection. J. Am. Vet. Med. Assoc., 1975, 166:993-995.

1595. Mengeling, W.L., Gutekunst, D.E., Pirtle, E.C. and Paul, P.S. Immunogenicity of bivalent vaccine for reproductive failure of swine induced by pseudorabies virus and porcine parvovirus. Am. J. Vet. Res., 1981, 42:600-603.

1596. Mengeling, W.L. and Paul, P.S. Reproductive performance of gilts exposed to porcine parvovirus at 56 or 70 days of gestation. Am. J. Vet. Res., 1981, 42:2074-2076.

1597. Mengeling, W.L., Paul, P.S. and Brown, T.T. Transplacental infection and embryonic death following maternal exposure to porcine parvovirus near the time of conception. Arch. Virol., 1980, 65:55-62.

1598. Mengeling, W.L., Pejsak, Z. and Paul, P.S. Biological assay of attenuated strain NADL-2 and virulent strain NADL-8 or porcine parvovirus. Am. J. Vet. Res., 1984, 45:2403-2407.

1599. Menser, M.A., Dods, C. and Harley, J.D. A 25-year follow-up of congenital rubella. Lancet, 1967a, 2:1347-1350.

1600. Menser, M.A. and Forrest, J.M. Maternal infections and the developing foetus. Med. J. Austr., 1973, 1:448.

1601. Menser, M.A. and Forrest, J.M. Rubella vaccination and pregnancy. Med. J. Aust., 1975, 1:794-795.

1602. Menser, M.A., Harley, J.D., Hertzberg, R., Dorman, D.C. and Murphy, A.M. Persistence of virus in lens for three years after prenatal rubella. Lancet, 1967, 2:387.

1603. Menser, M.A., Harley, J.D., Housego, C.J. and Murphy, A.M. Possible chemical factors in the postnatal development of rubella cataracts. Lancet, 1966, 2:771-772.

1604. Menser, M.A. and Reye, R.D.K. The pathology of congenital rubella: a review written by request. Pathology, 1974, 6:215.

1605. Menser, M.A., Robertson, S.E.J., Dorman, D.C., Gillespie, A.M. and Murphy, A.M. Renal lesions in congenital rubella. Pediatrics, 1967c, 40:901-904.

1606. Mensik, J. and Pokorny, J. Development of antibody response to swine influenza virus in pigs. I. The influence of experimental infection of pregnant sows on serum antibody production by their progeny during postnatal development. Zentralbl. Veterinarmed. [B], 1971, 18:177-189.

1607. Metz, H.S. and Harkey, M.E. Pigmentary retinopathy following maternal measles (morbilli) infection. Am. J. Ophthalmol., 1968, 66:1107-1110.

1608. Meurman, O., Terho, P. and Salmi, A. Activation of rheumatoid factor during pregnancy. Lancet, 1978, 2:685-686.

1609. Meyer, H., Liess, B., Frey, H.R., Hermanns, W. and Trautwein, G. Experimental transplacental transmission of hog cholera virus in pigs. IV. Virological and serological studies in newborn piglets. Zentralbl. Veterinarmed. [B], 1981, 28:659-668.

1610. Meyer, J. and Loffler, H. Infection of gravia mice with Coxsackievirus. Path. Microbiol., 1968, 32:139.

1611. Meyers, J.D. Congenital varicella in term infants: risk reconsidered. J. Infect. Dis., 1974, 129:215-217.

1612. Michaels, R.H. and Kenny, F.M. Postnatal growth retardation in congenital rubella. Pediatrics, 1969, 43:251-259.

1613. Mihailescu, R. and Petrovici, M. Effect of smallpox vaccination on the product of conception. Arch. Roum. Pathol. Exp. Microbiol., 1975, 34:67-74.

1614. Mikkelsen, M., Skinhoj, P. and Olesen, H. Cytogenetical findings in children born to carriers of hepatitis-associated antigen. Acta. Pathol. Microbiol. Scand. [B], 1973, 81:207-218.

1615. Miller, C.G. and Thorburn, M.J. An outbreak of congenital rubella in Jamaica. Clinical features. West Indian Med. J., 1966, 15:177-188.

1616. Miller, C.L., Miller, E., Sequeira, P.J., Cradock-Watson, J.E., Longson, M. and Wiseberg, E.C. Effect of selective vaccination on rubella susceptibility and infection in pregnancy. Br. Med. J. [Clin. Res.], 1985, 291:1398-1401.

1617. Miller, E., Cradock-Watson, J.E. and Pollock, T.M. Consequences of confirmed maternal rubella at successive stages of pregnancy. Lancet, 1982, 2:781-784.

1618. Miller, J.A. and Hermon, Y.E. Hepatitis B in pregnant women. NZ. Med. J., 1984, 97:461.

1619. Miller, J.M. and VanDerMaaten, M.J. Experimentally induced infectious bovine rhinotracheitis virus infection during early pregnancy: effect on the bovine corpus luteum and conceptus. Am. J. Vet. Res., 1986, 47:223-228.

1620. Miller, M.H., Rabinowitz, M. and Cohen, M. Pure-tone audiometry in prenatal rubella. Arch. Otolaryngol., 1971, 94:25-29.

1621. Miller, M.H., Rabinowitz, M., Frost, J.O. and Seager, G.M. Audiological problems associated with maternal rubella. Laryngoscope, 1969, 79:417-426.

1622. Miller, M.J., Seaman, E. and Remington, J.S. Clinical spectrum of congenital toxoplasmosis. Problems in recognition. J. Pediat., 1967, 70:714.

1623. Miller, W.C. Equine rhinopneumonitis virus as a cause of abortion in pregnant mares. Proc. R. Soc. Med., 1966, 59:1080-1081.

1624. Miller, Z.B. Chromosome abnormalities in measles. Lancet, 1963, 2:1070.

1625. Mills, C. Deafness and rubella: a challenge and a charge to rehabilitation. Am. Ann. Deaf., 1980, 125:1026-1030.

1626. Milne, A. Viral hepatitis in the eastern Bay of Plenty. NZ. Med. J., 1980, 92:87-91.

1627. Milne, A. Transplacental transmission of hepatitis B virus. Lancet, 1986, 1:860-861.

1628. Milner, A.R. and Marshall, I.D. Pathogenesis of in utero infections with abortogenic and non-abortogenic alphaviruses in mice. J. Virol., 1984, 50:66-72.

1629. Milner, A.R., Marshall, I.D. and Mullbacher, A. Effect of pregnancy on stimulation of alphavirus immunity in mice. J. Virol., 1984, 50:73-76.

1630. Mims, C.A. Pathogenesis of viral infections of the fetus. Prog. Med. Virol., 1968, 10:194-237.

1631. Mims, C.A. Effect on the fetus of maternal infection with lymphocytic choriomeningitis (LCM) virus. J. Infect. Dis., 1969, 120:582-597.

1632. Minamishima, Y., Graham, B.J. and Benyesh-Melnick, M. Neutralizing antibodies to cytomegaloviruses in normal simian and human sera. Infect. Immun., 1971, 4:368-373.

1633. Minkoff, H. Hepatitis. Clin. Obstet. Gynecol., 1983, 26:178-185.

1634. Minkoff, H. Acquired immunodeficiency syndrome. J. Nurse Midwifery, 1986, 31:189-193.

1635. Minkoff, H.L. and Schwartz, R.H. AIDS: time for obstetricians to get involved. Obstet. Gynecol., 1986, 68:267-268.

1636. Minnefor, A., Cooper, R., Thomas, K., DeLaCruz, A., Ahdieh, H., Guerrero, I., Joshi, V.V. and Desposito, F. Immune deficiency syndrome in children. Journal of the American Medical Association, 1983, 249:2345-2349.

1637. Mirand, E.A. Murine viral-induced polycyhemia. Ann. NY. Acad. Sci., 1968, 149, 486-496.

1638. Mirand, E.A., Buffett, R.F. and Grace, J.T. Jr. Mode of transmission of Friend virus disease. Proc. Soc. Exp. Biol. Med., 1966, 121:970-976.

1639. Mirand, E.A., Grace, J.T. and Buffett, R.F. Passive and active immunity to Friend virus disease. Nature, 1966, 209:696-698.

1640. Mirchamsy, H., Shafyi, A., Mahinpour, M. and Nazari, P. Age of measles immunization in tropics. Dev. Biol. Stand., 1978, 41:191-194.

1641. Mitchell, D. Vaccines for EHV1 abortion. Vet. Rec., 1983, 112:285.

1642. Mitchell, D. and Greig, A.S. The incidence and significance of bovine herpesvirus (infectious bovine rhinotracheitis) antibodies in the sera of aborting cattle. Can. J. Comp. Med., 1967, 31:234-238.

1643. Mittelmann-Handwerker, S., Pardes, J.G., Post, R.C., Sumala, M., Rosenberg, J.C. and Chervenak, F.A. Fetal ventriculomegaly and brain atrophy in a woman with intrauterine cytomegalovirus infection. A case report. J. Reprod. Med., 1986, 31:1061-1064.

1644. Miura, T., Sugamata, M., Ogata, T. and Matsuda, R. Japanese encephalitis virus infection in fetal mice at different stages of pregnancy. II. Resistance to Japanese encephalitis virus infection. Acta. Virol. (Praha.), 1982, 26:283-287.

1645. Miura, Y., Hayashi, S., Ishihara, T., Inaba, Y. and Omori,, T. Neutralizing antibody against Akabane virus in precolostral sera from calves with congenital arthrogryposis-hydranencephaly syndrome. Arch. Gesamte. Virusforsch., 1974, 46:377-380.

1646. Miyoshi, I., Hasegawa, H., Tsubota, T. and Hiraki, K. Simultaneous determination of antibody to Epstein-Barr virus in prenatal mothers and new-born infants. Experientia, 1972, 28:195-196.

1647. Modlin, J.F. Fatal ECHOvirus 11 disease in premature neonates. Pediatrics, 1980, 66:775-780.

1648. Modlin, J.F., Brandling-Bennett, A.D., Witte, J.J., Campbell, C.C. and Meyers, J.D. A review of five years' experience with rubella vaccine in the United States. Pediatrics, 1975, 55:20-29.

1649. Modlin, J.F. and Crumpacker, C.S. Coxsackievirus B infection in pregnant mice and transplacental infection of the fetus. Infect. Immun., 1982, 37:222-226.

1650. Modlin, J.F., Herrmann, K., Brandling-Bennett, A.D., Eddins, D.L. and Hoyden, G.F. Risk of congenital abnormality after inadvertent rubella vaccination of pregnant women. N. Engl. J. Med., 1976, 294:972-974.

1651. Modlin, J.F., Polk, B.F., Horton, P., Etkind, P., Crane, E. and Spiliotes, A. Perinatal ECHOvirus infection: risk of transmission during a community outbreak. N. Engl. J. Med., 1981, 305:368-371.

1652. Moe, J.B. and Osburn, B.I. Fetal and maternal immunologic manifestations of intrauterine Adenovirus SV-20 infection. Vet. Immunol. Immunopathol., 1983, 4:319-332.

1653. Moe, J.B., Osburn, B.I., Schwartz, L.W., Anderson, J. and Johnson, K.P. Experimental adenovirus SV-20 pneumonia in fetal rhesus monkeys. Pathologic and virologic studies. Lab. Invest., 1979, 41:211-219.

1654. Mohan, M., Mehta, P.K., Sehgal, S., Prabhakar, A.K. and Bhargava, S.K. Optimum age of measles immunization -- maternal and transplacentally transmitted measles antibodies in infancy. Indian Pediatr., 1981, 18:631-635.

1655. Molello, J.A., Chow, T.L., Owen, N. and Jensen, R. Placental pathology. V. Placental lesions of cattle experimentally infected with infectious bovine rhinotracheitis virus. Am. J. Vet. Res., 1966, 27:907-915.

1656. Mollica, F., Musumeci, S. and Fischer, A. Neonatal hepatitis in five children of a hepatitis B surface antigen carrier woman. J. Pediatr., 1977, 90:949-951.

1657. Mollica, F., Musumeli, S., Rugolo, S. and Mattina, T. A prospective study of 18 infants of chronic HBsAg mothers. Arch. Dis. Child, 1979, 54:750-754.

1658. Mollica, F., Pavone, L., Nuciforo, G. and Sorge, G. A case of cyclopia. Role of environmental factors. Clin. Genet., 1979, 16:69-71.

1659. Molnarova, A. and Blaskovic, D. The influence of influenza virus infection of pregnant mice on the development of their foetuses. Acta. Virol. (Praha.), 1975, 19:259.

1660. Molnarova, A. and Blaskovic, D. The influence of influenza infection on praenatal development of mice. Acta. Univ. Carol. [Med.] (Praha.), 1978, 24:43-55.

1661. Molnarova, A. and Mayer, V. Experimental infection of pregnant mice with viruses of the tick-borne encephalitis (TBE) complex. Acta. Virol. (Praha.), 1980, 24:297.

1662. Mondal, H. and Hofschneider, P.H. Isolation and characterization of retrovirus-like elements from normal human fetuses. Int. J. Cancer, 1982, 30:281-287.

1663. Money, T.D. Exposure to rubella in pregnancy. Br. Med. J., 1972, 4:666-667.

1664. Monif, G.R. Rubella vaccination. An evolving problem for obstetrics and gynecology. Obstet. Gynecol., 1972, 39:304-307.

1665. Monif, G.R. Maternal mumps infection during gestation: observations on the progeny. Am. J. Obstet. Gynecol., 1974, 119:549-551.

1666. Monif, G.R. Rubella virus and the pancreas. Med. Chir. Dig., 1974, 3:195-197.

1667. Monif, G.R. Viruses as teratogens. Clin. Obstet. Gynecol., 1975, 18:209-220.

1668. Monif, G.R. and Dische, R.M. Viral placentitis in congenital cytomegalovirus infection. Am. J. Clin. Pathol., 1972, 58:445-449.

1669. Monif, G.R., Egan, E.A. II, Held, B. and Eitzman, D.V. The correlation of maternal cytomegalovirus infection during varying stages in gestation with neonatal involvement. J. Pediatr., 1972, 80, 17-20.

1670. Monif, G.R. and Hardt, N.S. Management of herpetic vulvovaginitis in pregnancy. Semin. Perinatol., 1983, 7:16-21.

1671. Monif, G.R., Hardy, J.H. and Sever, J.L. The lack of association between the appearance of complement fixing antibodies and the recovery of virus in a child with congenital rubella. Pediatrics, 1967, 39:289-291.

1672. Monif, G.R. and Jordan, P.A. Rubella virus and rubella vaccine. Semin. Perinatol., 1977, 1:41-49.

1673. Monif, G.R., Kellner, K.R. and Donnelly, W.H. Jr. Congenital herpes simplex type II infection. Am. J. Obstet. Gynecol., 1985, 152:1000-1002.

1674. Monif, G.R. and Sever, J.L. Chronic infection of the central nervous system with rubella virus. Neurology, 1966, 16:111-112.

1675. Monif, G.R. and Sever, J.L. Pathogenesis of the teratogenic effect of rubella virus. Obstet. Gynecol., 1966, 27:741-743.

1676. Monif, G.R.G., Sever, J.L., Schiff, G.M. and Traub, R.G. Isolation of rubella virus from products of conception. Amer. J. Obst. Gyn., 1965, 91:1143.

1677. Monif, G.R., Sowards, D.L. and Eitzman, D.V. Serologic and immunologic evaluation of neonates following maternal influenza infection during the second and third trimesters of gestation. Am. J. Obstet. Gynecol., 1972, 114:239-242.

1678. Monif, R.G., Egan, E.A., Held, B. and Eitzman, D.V. The correlation of maternal cytomegalovirus infection during varying stages in gestation with neonatal involvement. J. Pediatr., 1972, 80:17-20.

1679. Monitor. Pregnancy hazard of influenza virus. New Scientist, 1975, 5:515.

1680. Montali, R.J., Allen, G.P., Bryans, J.T., Phillips, L.G. and Bush, M. Equine herpesvirus type 1 abortion in an onager and suspected herpesvirus myelitis in a zebra. J. Am. Vet. Med. Assoc., 1985, 187:1248-1249.

1681. Montgomery, J.R., Flanders, R.W. and Yow, M.D. Congenital anomalies and herpesvirus infection. Am. J. Dis. Child, 1973, 126:364-366.

1682. Montgomery, R., Youngblood, L. and Medearis, D.N. Jr. Recovery of cytomegalovirus from the cervix in pregnancy. Pediatrics, 1972, 49:524-531.

1683. Moqabqab, W.J. and Stowe, F.R. Jr. Evaluation of attenuated rubella virus vaccine in families. Am. J. Dis. Child, 1971, 122:122-128.

1684. Mordhorst, C.H. and Dawson, C. Sequelae of neonatal inclusion conjunctivitis and associated disease in parents. Am. J. Ophthalmol., 1971, 71:861-867.

1685. Moreland, A.F., Gaskin, J.M., Schimpff, R.D., Woodard, J.C. and Olson, G.A. Effects of influenza, mumps, and western equine encephalitis viruses on fetal rhesus monkeys (Macaca mulatta). Teratology, 1979, 20:53-64.

1686. Moreland, A.F., Schimpff, R.D. and Gaskin, J.M. Fetal mortality and malformations associated with experimental infections of western equine encephalomyelitis vaccine virus in rhesus monkeys (Macaca mulatta). Teratology, 1979, 20:65-74.

1687. Morgan-Capner, P., Hodgson, J., Hambling, M.H., Dulake, C., Coleman, T.J., Boswell, P.A., Watkins, R.P., Booth, J., Stern, H., Best, J.M. et al. Detection of rubella-specific IgM in subclinical rubella reinfection in pregnancy. Lancet, 1985, 1:244-246.

1688. Morgan-Capner, P., Rodeck, C.H., Nicolaides, K.H. and Cradock-Watson, J.E. Prenatal detection of rubella-specific IgM in fetal sera. Prenat. Diagn., 1985, 5:21-26.

1689. Morimoto, T., Fujisaki, Y., Ito, Y. and Tanaka, Y. Biological and physicochemical properties of porcine parvovirus recovered from stillborn piglets. Natl. Inst. Anim. Health Q. (Tokyo), 1972, 12:137-144.

1690. Morimoto, T., Kurogi, H., Miura, Y., Sugimori, T. and Fujisaki, Y. Isolation of Japanese encephalitis virus and a hemagglutinating DNA virus from the brain of stillborn piglets. Natl. Inst. Anim. Health Q. (Tokyo), 1972, 12:127-136.

1691. Moroni, G.A., Costantino, D., Zampieri, G., Gianotti, G.A., Doglia, M. and DelPrete, S. Do the hepatitis antigens cross the placenta? Lancet, 1971, 2:376.

1692. Moroso, M.J. Genetic disposition to deafness in maternal rubella: fact or myth? J. Otolaryngol., 1985, 14:44-48.

1693. Morrison, R.B. and Joo, H.S. Prenatal and preweaning deaths caused by pseudorabies virus and porcine parvovirus in a swine herd. J. Am. Vet. Med. Assoc., 1985, 187:481-483.

1694. Morse, H.C. III, Yetter, R.A., Stimpfling, J.H., Pitts, O.M., Fredrickson, T.N. and Hartley, J.W. Greying with age in mice: relation to expression of murine leukemia viruses. Cell, 1985, 41:439-448.

1695. Mortimer, P.P., Cohen, B.J., Buckley, M.M., Cradock-Watson, J.E., Ridehalgh, M.K., Burkhardt, F. and Schilt, U. Human parvovirus and the fetus. Lancet, 1985, 2:1012.

1696. Morton, R., Mitchell, I and McNicol, E. Neonatal cytomegalic inclusion disease in a set of twins one member of whom was a hydropic stillbirth the other completely uninfected. Case report. Br. J. Obstet. Gynaecol., 1983, 90:276-279.

1697. Moscari, E., Kudron, E., Szalay, D., Horvath, I., Szabo, L. and Toplak, L. Isolation of porcine parvovirus (PPV) from swine herds affected by reproductive failure, and serologic evidence of infection in Hungarian large swine herds. Acta. Vet. Hung., 1983, 31:5-15.

1698. Mosley, J.W. Vertical transmission of type B hepatitis. J.A.M.A., 1972, 220:1128-1129.

1699. Mosley, J.W. Viral hepatitis: a group of epidemiologic entities. Can. Med. Assoc. J., 1972, 106:427-434.

1700. Moss, P.D., Heffernan, C.K., Thurston, J.G. and Robertson, L. Enteroviruses and congenital abnormalities. Br. Med. J., 1967, 2:110-111.

1701. Mostoufi-Zadeh, M., Driscoll, S.G., Biano, S.A. and Kundsin, R.B. Placental evidence of cytomegalovirus infection of the fetus and neonate. Arch. Pathol. Lab. Med., 1984, 108:403-406.

1702. Mostoufi-Zadeh, M. Lack, E.E., Gang, D.L., Perez-Atayde, A.R. and Driscoll, S.G. Postmortem manifestations of ECHO virus 11 sepsis in five newborn infants. Hum. Pathol., 1983, 14:818-823.

1703. Muir, B. Jr. Maternal and congenital rubella -- past, present and future. Med. Ann. DC., 1970, 39:243-248.

1704. Mukherjee, P., Sengupta, K.P. and Mukherjee, K.L. Passage of poliovirus antibodies across the placental barrier. Indian J. Med. Res., 1980, 71:840-846.

1705. Mullooly, J.P., Barker, W.H. and Nolan, T.F. Jr. Risk of acute respiratory disease among pregnant women during influenza A epidemics. Public Health Rep., 1986, 101:205-211.

1706. Munoz, N. Perinatal viral infections and the risk of certain cancers. Prog. Biochem. Pharmacol., 1978, 14:104-108.

1707. Murasko, D.M. and Blank, K.J. Transplacental interferon-mediated effect in AKR mice. Virology, 1980, 106:148-151.

1708. Murata, S., Nomoto, K., Hattori, T., Iami, M. and Miyakawa, Y. Maternal transmission of hepatitis B. N. Engl. J. Med., 1977, 296:692-693.

1709. Murphy, W.F. and Flannery, J.L. Congenital toxoplasmosis occurring in identical twins. Amer. J. Dis. Child, 1952, 84:223-226.

1710. Murray, D.L., Imagawa, D.T., Okada, D.M. and St. Geme, J.W. Jr. Antibody response to monovalent A/New Jersey/8/76 influenza vaccine in pregnant women. J. Clin. Microbiol., 1979, 10:184-187.

1711. Murthy, V.K., Sawhney, I.M., Prabhakar, S. and Chopra, J.S. Isolated facial palsy in chickenpox. J. Neurol. Neurosurg. Psychiatry, 1984, 47:754.

1712. Music, S.I., Fine, E.M. and Togo, Y. Zoster-like disease in newborn due to herpes simplex virus. New Eng. J. Med., 1971, 284:24.

1713. Mutch, M.G. and Barlow, J.F. Twenty eight year old woman with term pregnancy and jaundice. SD. J. Med., 1975, 28:9-13.

# N

1714. Naeye, R.L. Cytomegalic inclusion disease. The fetal disorder. Amer. J. Clin. Path., 1967, 47:738.

1715. Naeye, R.L. Infants of prolonged gestation. A necropsy study. Arch. Path., 1967, 84:37.

1716. Naeye, R.L. Fetal growth with congenital syphilis: a quantitative study. Amer. J. Clin. Path., 1971, 55:228.

1717. Naeye, R.L. and Blanc, W. Pathogenesis of congenital rubella. J.A.M.A., 1965, 194:1277-1283.

1718. Naggan, L., Bar-Shany, S., Shmuelewitz, O., Finzi, M. and Prywes, R. Prevalence of hepatitis B markers (HBsAg and HBsAb) in women screened at a time of delivery. Isr. J. Med. Sci., 1980, 16:347-350.

1719. Nagaratnam, A., Alagaratnam, K. and Thambapillai, N.J. Recurrent intrahepatic cholestatic jaundice of pregnancy complicated by viral hepatitis. Ceylon Med. J., 1976, 21:25-28.

1720. Naggan, L., Gorodischer, R., Bar-Shany, S. and Sarov, B. Clinical hepatitis B in two infants born to asymptomatic hepatitis B surface antigen carrier mothers. Isr. J. Med. Sci., 1983, 19:292-293.

1721. Nahmias, A. Perinatal risk associated with maternal genital herpes simplex virus infection. Am. J. Obstet. Gynecol., 1971, 110:835-837.

1722. Nahmias, A.J., Alford, C.A. and Korones, S.B. Infection of the newborn with herpesvirus hominis. Adv. Pediatr., 1970, 17:185-226.

1723. Nahmias, A.J. and Hagler, W.S. Ocular manifestation of herpes simplex in the newborn (neonatal ocular herpes). Int. Ophthalmol. Clin., 1972, 12:191-213.

1724. Nahmias, A.J., Josey, W.E. and Naib, Z.M. Neonatal herpes simplex infection. Role of genital infection in mother as the source of virus in the newborn. J.A.M.A., 1967, 199:164-168.

1725. Nahmias, A.J., Josey, W.E. and Naib, Z.M. Significance of herpes simplex virus infection during pregnancy. Clin. Obstet. Gynecol., 1972, 15:929-938.

1726. Nahmias, A.J., Josey, W.E., Naib, Z.M., Freeman, M.G., Fernandez, R.J. and Wheeler, J.H. Perinatal risk associated with maternal genital herpes simplex virus infection. Am. J. Obstet. Gynecol., 1971, 110:825-837.

1727. Nahmias, A.J., Naib, Z.M., Josey, W.E. and Clepper, A.C. Genital herpes simplex infection. Virologic and cytologic studies. Obstet. Gynecol., 1967, 29:395-400.

1728. Nahmias, A.J. and Schwahn, M.G. Neonatal herpes simplex: a worldwide disease which is potentially preventable and treatable. Prog. Clin. Biol. Res., 1985, 163:355-362.

1729. Nahmias, A.J., Visintine, A.M. and Josey, W.E. Cesarean section and genital herpes. N. Engl. J. Med., 1977, 296:1359.

1730. Nahmias, A.J., Visintine, A.M., Reimer, C.B., DelBuono, I., Shore, S.L. and Starr, S.E. Herpes simplex virus infection of the fetus and newborn. In: Krugman, S., Gershon, A.A., ed. Infections of the fetus and the newborn infant. New York, Liss, 1975, 63-77.

1731. Nahum, L.H. Teratogenesis and viral infections in pregnancy. Conn. Med., 1968, 32:267-269.

1732. Naib, Z.M., Nahmias, A.J., Josey, W.E. and Wheeler, J.H. Association of maternal genital herpetic infection with spontaneous abortion. Obstet. Gynecol., 1970, 35:260-263.

1733. Nair, P.V., Weissman, J.Y., Tong, M.J., Thursby, M.W., Paul, R.H. and Henneman, C.E. Efficacy of hepatitis B immune globulin in prevention of perinatal transmission of the hepatitis B virus. Gastroenterology, 1984, 87:293-298.

1734. Naish, J. Rubella immunity in pregnant women in a north London practice. Br. Med. J. [Clin.] [Res.], 1984, 288:1053-1054.

1735. Nandapalan, N., Taylor, C.E., Greenwell, J., Scott, M., Scott, R., Hey, E.N. and Toms, G.L. Seasonal variations in maternal serum and mammary immunity to RS virus. J. Med. Virol., 1986, 20:79-87.

1736. Nandi, S. and Helmich, C. Transmission of mammary tumor virus by the GR mouse strain. I. Role of the virus in the production of lesions. J. Natl. Cancer Inst., 1974, 52:1285-1290.

1737. Nandi, S. and Helmich, C. Transmission of the mammary tumor virus by the GR mouse strain. II. Genetic studies. J. Natl. Cancer Inst., 1974, 52:1567-1570.

1738. Narita, M., Imada, T. and Fukusho, A. Pathologic changes caused by transplacental infection with an adenovirus-like agent in pigs. Am. J. Vet. Res., 1985, 46:1126-1129.

1739. Narita, M., Inui, S., Kawakami, Y., Kitamura, K. and Maeda, A. Histopathological changes of the brain in swine fetuses naturally infected with procine parvovirus. Natl. Inst. Anim. Health Q. (Tokyo), 1975, 15:24-28.

1740. Narwekar, M.R., Penkar, S.J., Pinto, S. and Jaffar, H. Jaundice in pregnancy (a case report of infective hepatitis in pregnancy with hypofibrinogenenia and hypoglycemic coma. J. Postgrad. Med., 1974, 20:196-199.

1741. Nastac, E., Stoian, M., Hozoc, M. and Athanasiu, P. Transplacental cancerization by Rous sarcoma virus. Virologie, 1979, 30:189-191.

1742. Neifeld, H. Intra-amniotic administration of prostaglandin E2 in midtrimester abortions. S. Afr. Med. J., 1974, 48:2614.

1743. Neighbour, P.A. The effect of maternal cytomegalovirus infection on preimplantation development in the mouse. J. Reprod. Fertil., 1976, 48:83-89.

1744. Neitz, W.O. Wesselsbron disease. Bull. Off. Int. Epizoot., 1966, 65:1735-1741.

1745. Nesburn, A.B. Progress against rubella. Arch. Ophthalmol., 1967, 77:425-426.

1746. Nettleton, P.F., Herring, J.A., Hogg, R.A. and Nicolson, T.B. Laboratory confirmation of IBR virus induced abortion. Vet. Rec., 1981, 108:243.

1747. Neufeld, J.L. Spontaneous pustular dermatitis in a newborn piglet associated with a poxvirus. Can. Vet. J., 1981, 22:156-158.

1748. Newburger, J.W., Silbert, A.R., Buckley, L.P. and Fyler, D.C. Cognitive function and age at repair of transposition of the great arteries in children. New England Journal of Medicine, 1984, 310:1495-1499.

1749. Newman, S.J., McCallin, P.F. and Sever, J.L. Attempts to isolate H-1 virus from spontaneous human abortions: a negative report. Teratology, 1970, 3:279-281.

1750. Nichols, W.W. Role of viruses in the etiology of chromosomal abnormalities. Amer. J. Human Genet., 1966, 18:81.

1751. Nieburg, P. and Gardner, L. Thyroiditis and congenital rubella syndrome. J. Pediatr., 1976, 89:156.

1752. Nielson, S.L., Urmacher, C. and Petito, C.K. Encephalopathy in acquired immune deficiency syndrome (AIDS). Journal of Neuropathology and Experimental Neurology, 1983, 42:347.

1753. Niemi, S.M., Evermann, J.F., Huffman, E.M. and Kirk, J.H. Border disease virus isolation from postpartum ewes. Am. J. Vet. Res., 1982, 43:86-88.

1754. Norden, C.W. and Kass, E.H. Bacteriuria of pregnancy - a critical appraisal. Ann. Rev. Med., 1969, 19:431.

1755. Northrop, R.L., Gardner, W.M. and Geittmann, W.F. Rubella reinfection during early pregnancy: a case report. Obstet. Gynecol., 1972, 39:524-526.

1756. Nouasria, B., Auoati, A., Bernau, J. et al. Fulminant viral hepatitis and pregnancy in Algeria and France. Ann. Trop. Med. Parasitol., 1986, 80:623-629.

1757. Numazaki, Y., Yano, N., Morizuka, T., Takai, S. and Ishida, N. Primary infection with human cytomegalovirus: virus isolation from healthy infants and pregnant women. Am. J. Epidemiol., 1970, 91:410.

1758. Nusbacher, J., Hirschhorn, K. and Cooper, L.Z. Chromosomal abnormalities in congenital rubella. N. Engl. J. Med., 1967, 276:1409-1413.

# O

1759. Obayashi, A. and Mayumi, M. HBAg subtypes in familial carriers. Lancet, 1974, 1:507.

1760. Obenour, L.C. The electroretinogram in rubella retinopathy. Int. Ophthalmol. Clin., 1972, 12:105-110.

1761. Odessky, L., Newman, B. and Wein, G.B. Congenital varicella. N.Y. State J. Med., 1954, 54:2849.

1762. O'Donovan, J.D. Bell's palsy in pregnancy. Lancet, 1973, 2:667.

1763. Ogata, M., Nagao, Y., Jitsunari, F., Kitamura, N. and Okazaki, T. Infection of herons and domestic fowls with Japanese encephalitis virus with specific reference to maternal antibody of hen (epidemiological study on Japanese encephalitis 26). Acta. Med. Okayama, 1970, 24:175-184.

1764. Ogilvie, M.M., Stephens, J.R. and Larkin, M. Chickenpox in pregnancy. Lancet, 1986, 1:915-916.

1765. Ogilvie, M.M. and Tearne, C.F. Spontaneous abortion after hand-foot-and-mouth disease caused by Coxsackie virus A16. Br. Med. J., 1980, 281:1527-1528.

1766. Ogunbode, O. Jaundice during pregnancy in Ibadan, Nigeria. Int. J. Gynaecol. Obstet., 1978-79, 16:289-292.

1767. Ogunmekan, D.A., Bracken, P. and Marshall, W.C. An assessment of the effectiveness of the measles immunization programme in Lagos, Nigeria. Ann. Trop. Med. Parasitol., 1981, 75:87-92.

1768. Ohba, N. Formation of embryonic abnormalities of the mouse by a viral infection of mother animals. Acta. Path. Jpn., 1958, 8:127.

1769. Ohmann, H.B. Experimental fetal infection with bovine viral diarrhea virus. II. Morphological reactions and distribution of viral antigen. Can. J. Comp. Med., 1982, 46:363-369.

1770. Ohmann, H.B., Jensen, M.H., Srensen, K.J. and Dalsgaard, K. Experimental fetal infection with bovine viral diarrhea virus. I. Virological and serological studies. Can. J. Comp. Med., 1982, 46:357-362.

1771. Ohshima, K., Morimoto, N., Kagawa, Y., Numakunai, S., Hirano, T. and Kayano, H. A survey for maternal antibodies to bovine leukemia virus (BLV) in calves born to cows infected with BLV. Nippon. Juigaku. Zasshi., 1984, 46:583-586.

1772. Ohto, H., Lin, H.H., Kawana, T., Etoh, T. and Tohyama, H. Intrauterine transmission of hepatitis B virus is closely related to placental leakage. J. Med. Virol., 1987, 21:1-6.

1773. Ojala, P., Vesikari, T. and Elo, O. Rubella during pregnancy as a cause of congenital hearing loss. Am. J. Epidemiol., 1973, 98:395-401.

1774. Okada, K., Kamiyama, I., Inomata, M., Imai, M. and Miyakawa, Y. Antigen and anti-e in the serum of asymptomatic carrier mothers as indicators of positive and negative transmission of hepatitis B virus to their infants. N. Engl. J. Med., 1976, 294:746-749.

1775. Okey, C.H. Rubella serology -- The HAI test. J. Maine Med. Assoc., 1967, 58:49-50.

1776. Okochi, K. and Sato, H. Adult T-cell leukemia virus, blood donors and transfusion: experience in Japan. Prog. Clin. Biol. Res., 1985, 182:245-256.

1777. Oldstone, M.B. and Dixon, F.J. Pathogenesis of chronic disease associated with persistent lymphocytic choriomeningitis viral infection. II. Relationship of the anti-lymphocytic choriomeningitis immune response to tissue injury in chronic lymphocytic choriomeningitis disease. J. Exp. Med., 1970, 131:1-19.

1778. Oldstone, M.B. and Dixon, F.J. Persistent lymphocytic choriomeningitis viral infection. 3. Virus-anti-viral antibody complexes and associated chronic disease following transplacental infection. J. Immunol., 1970, 105:829-837.

1779. Oldstone, M.B. and Dixon, F.J. Disease accompanying in utero viral infection. The role of maternal antibody in tissue injury after transplacental infection with lymphocytic choriomeningitis virus. J. Exp. Med., 1972, 135:827-838.

1780. O'Neill, J.F. Strabismus in the rubella syndrome. Int. Ophthalmol. Clin., 1972, 12:111-120.

1781. O'Neill, R.T. and Schwartz, R.H. Cytomegaloviremia as a cause of delayed posttransfusion fever. Am. J. Obstet. Gynecol., 1970, 107:970-971.

1782. Onuma, M., Baumgartener, L.E., Olson, C. and Pearson, L.D. Fetal infection with bovine leukemia virus in sheep. Cancer Res., 1977, 37:4075-4081.

1783. Oppenheimer, E.H. and Esterly, J.R. Cytomegalovirus infection: A possible cause of biliary atresia. Amer. J. Path., 1973, 71:2a.

1784. Orban, S., Liess, B., Hafez, S.M., Frey, H.R., Blindow, H. and Sasse-Patzer, B. Studies on transplacental transmissibility of a Bovine Virus Diarrhoea (BVD) vaccine virus. I. Inoculation of pregnant cows 15 to 90 days before parturition (190th to 265th day of gestation). Zentralbl. Veterinarmed. [B], 1983, 30:619-634.

1785. O'Reilly, K.J., Paterson, J.S. and Harriss, S.T. The persistence in kittens of maternal antibody to feline infectious enteritis (panleukopenia). Vet. Rec., 1969, 84:376-378.

1786. Ornoy, A., Dudai, M. and Sadovsky, E. Placental and fetal pathology in infectious mononucleosis. A possible indicator for Epstein-Barr virus teratogenicity. Diagn. Gynecol. Obstet., 1982, 4:11-16.

1787. Ornoy, A., Segal, S., Nishmi, M., Simcha, A. and Polishuk, W.Z. Fetal and placental pathology in gestational rubella. Am. J. Obstet. Gynecol., 1973, 116:949-956.

1788. Orr, M.B., Barlow, R.M. and Ryder, M.L. Histological and histochemical studies of fetal skin follicles in Border disease of sheep. Res. Vet. Sci., 1977, 22:56-61.

1789. Ortaldo, J.R., Ting, C.C. and Herberman, R.B. Modulation of fetal antigen(s) in mouse leukemia cells. Cancer Res., 1974, 34:1366-1371.

1790. Osburn, B.I. The ontogeny of the ruminant immune system and its significance in the understanding of maternal-fetal-neonatal relationships. Adv. Exp. Med. Biol., 1981, 137:91-103.

1791. Osburn, B.I., Johnson, R.T., Silverstein, A.M., Prendergast, R.A., Jochim, M.M. and Levy, S.E. Experimental viral-induced congenital encephalopathies. II. The pathogenesis of bluetongue vaccine virus infection in fetal lambs. Lab. Invest., 1971, 25:206-210.

1792. O'Shaunessey, W.J. and Buechner, H.A. Hepatitis associated with a Coxsackie B5 virus infection during late pregnancy. J. Am. Med. Ass., 1962, 179:71.

1793. Osmundsen, P.E. Herpes gestationis. Report on 5 cases. Dermatologica, 1966, 132:393-399.

1794. Osterhammel, D. Audiological findings in children with a congenital rubella syndrome. Scand. Audiol., 1980, 33-44.

1795. Overall, J.C. Jr. Intrauterine virus infections and congenital heart disease. Am. Heart J., 1972, 84:823-833.

1796. Overall, J.C. Intra-uterine virus infections and congenital heart disease. Am. Heart J., 1972, 84:823.

1797. Overall, J.C. Jr. and Glasgow, L.A. Fetal response to viral infection: interferon production in sheep. Science, 1970, 167:1139-1141.

1798. Overall, J.C. Jr. and Glasgow, L.A. Virus infections of the fetus and newborn infant. J. Pediatr., 1970, 77:315-333.

1799. Overall, J.C. and Glasgow, L.A. Virus infection of the fetus and newborn infant. J. Pediat., 1970, 77:315.

1800. Owen, N.V., Chow, T.L. and Molello, J.A. Infectious bovine rhinotracheitis: correlation of fetal and placental lesions with viral isolations. Am. J. Vet. Res., 1968, 29:1959-1965.

1801. Owen, N.V., Chow, T.L. and Molello, J.A. Infectious bovine rhinotracheitis: relationship of level of maternal antibody titer to incidence of fetal death and abortion. Am. J. Vet. Res., 1968, 29:1967-1970.

1802. Oxford, J.S. and Schild, G.C. Growth of rubella virus in the hamster (mesocricetus auratus). Virology, 1966, 28:780.

1803. Oxford, J.S. and Sutton, R.N. The effect of rubella and herpesvirus homonis on the pre- and post-implantation stages of pregnancy in laboratory animals. J. Embryol. Exp. Morphol., 1968, 20:285-294.

1804. Oyrzanowska-Poplewska, J. Immunology of experimental rabies in rats: influence of the age factor. Acta. Microbiol. Pol. [A], 1969, 1:55-59.

# P

1805. Packer, A.D. The influence of maternal measles (morbilli) on the unborn child. Med. J. Austral., 1950, 1:835.

1806. Padgett, G.A., Gorham, J.R. and Henson, J.B. Epizootiologic studies of Aleutian disease. I. Transplacental transmission of the virus. J. Infect. Dis., 1967, 35-38.

1807. Palma, A.E. de. Rubella and the fetus. J.A.M.A., 1974, 227:324.

1808. Panizon, F., Rottini, G.D. and Tamaro, P. Au-anti-Au immunocomplexes in neonatal Au-associated hepatitis. Biol. Neonate, 1973, 22:274-279.

1809. Pantelick, E.L., Steere, A.C., Lewis, H.D. and Miller, D.J. Hepatitis B infection in hospital personnel during an eight-year period: policies for screening and pregnancy in high risk areas. Am. J. Med., 1981, 70:924-927.

1810. Papaevangelou, G. and Hoofnagle, J.H. Transmission of hepatitis B virus infection by asymptomatic chronic HBsAg carrier mothers. Pediatrics, 1979, 63:602-605.

1811. Papaevangelou, G., Hoofnagle, J. and Kremastinou, J. Transplacental transmission of hepatitis-B virus by symptom-free chronic carrier mothers. Lancet, 1974, 2:746-748.

1812. Papaioannou, A.C., Asrow, G.G. and Schuckmell, N.H. Nephrotic syndrome in early infancy as a manifestation of congenital syphilis. Pediatrics, 1961, 27:636.

1813. Pape, J.W., Liautaud, B., Thomas, F., Mathurin, J.R., St. Amand, M.M.A., Boncy, M., Pean, V., Pamphile, M., Laroche, A.C. and Johnson, W.D. Characteristics of the acquired immunodeficiency syndrome (AIDS) in Haiti. New England Journal of Medicine, 1983, 309:945-949.

1814. Parker, B.N., Wrathall, A.E. and Cartwright, S.F. Accidental introduction of porcine parvovirus and Talfan virus into a group of minimal disease gilts and their effects on reproduction. Br. Vet. J., 1981, 137:262-267.

1815. Parker, J., Plowright, W. and Pierce, M.A. The epizootiology of African swine fever in Africa. Vet. Rec., 1969, 85:668-674.

1816. Parker, P.L. Infectious hepatitis in pregnancy. Med. J. Aust., 1967, 2:967-970.

1817. Parkman, P.D., Phillips, P.E., Kirschstein, R.L. and Meyer, H.M. Jr. Experimental rubella virus infection in the rhesus monkey. J. Immunol., 1965, 95:743-752.

1818. Parkman, P.D., Phillips, P.E. and Meyer, H.M. Jr. Experimental rubella virus infection in pregnant monkeys. Am. J. Dis. Child, 1965, 110:390-394.

1819. Parrott, R.H., Kim, H.W., Brandt, C.D. and Chanock, R.M. Potential of attenuated respiratory syncytial virus vaccine for infants and children. Dev. Biol. Stand., 1975, 28:389-399.

1820. Parsonson, I.M., Della-Porta, A.J. and Snowdon, W.A. Congenital abnormalities in newborn lambs after infection of pregnant sheep with Akabane virus. Infect. Immun., 1977, 15:254-262.

1821. Parsonson, I.M., Della-Porta, A.J. and Snowdon, W.A. Developmental disorders of the fetus in some arthropod-borne virus infections. Am. J. Trop. Med. Hyg., 1981, 30:660-673.

1822. Parsonson, I.M., Della-Porta, A.B., Snowdon, W.A. and Murray, M.D. Congenital abnormalities in foetal lambs after inoculation of pregnant ewes with Akabane virus. Aust. Vet. J., 1975, 51:585-586.

1823. Parsonson, I.M. and McPhee, D.A. Bunyavirus pathogenesis. Adv. Virus Res., 1985, 30:279-316.

1824. Parsonson, I.M. and Snowdon, W.A. Experimental infection of pregnant cattle with ephemeral fever virus. Aust. Vet. J., 1974, 50:335-337.

1825. Partridge, J.W., Flewett, T.H. and Whitehead, J.E.M. Congenital rubella affecting an infant whose mother had rubella antibodies before conception. Br. Med. J., 1981, 282:187-188.

1826. Parving, A. and Starup, K. Audiometric and ophthalmological findings in rubella deafness. Audiology, 1977, 16:49-54.

1827. Paryani, S.G. and Arvin, A.M. Intrauterine infection with varicella-zoster virus after maternal varicella. N. Engl. J. Med., 1986, 314:1542-1546.

1828. Pass, R.F., Stagno, S., Dworsky, M.E., Smith, R.J. and Alford, C.A. Excretion of cytomegalovirus in mothers: observations after delivery of congenitally infected and normal infants. J. Infect. Dis., 1982, 146:1-6.

1829. Pass, R.F., Stagno, S., Myers, G.J. et al. Outcome of symptomatic congenital cytomegalovirus infection: results of long-term longitudinal follow-up. Pediatrics, 1980, 66:758-762.

1830. Paterson, W.G. and Smith, I.W. An unusual acute abdomen in pregnancy. An ECHO 8 virus infection. Practitioner, 1969, 203:337-339.

1831. Patrick, D.G. Report of a survey of children born in 1941 with reference to congenital abnormalities arising from maternal rubella. Med. J. Aust., 1948, 1:421.

1832. Paul, P.S. and Mengeling, W.L. Evaluation of a modified live-virus vaccine for the prevention of porcine parvovirus-induced reproductive disease in swine. Am. J. Vet. Res., 1980, 41:2007-2011.

1833. Pavel, A., Tirsia, E., Major, E. and Cristea, A. Detrimental effects of hepatitis B virus infection on the development of the product of conception. Virologie, 1983, 34:35-40.

1834. Payne, L.N. and Rennie, M. Pathogenesis of Marek's disease in chicks with and without maternal antibody. J. Natl. Cancer Inst., 1973, 51:1559-1573.

1835. Peacock, G.V. Biological requirements and control of equine rhinopneumonitis vaccine (live virus). J. Am. Vet. Med. Assoc., 1969, 155:310-312.

1836. Peacock, J.E. Jr. and Sarubbi, F.A. Disseminated Herpes simplex virus infection during pregnancy. Obstet. Gynecol., 1983, 61:13S-18S.

1837. Pearce, H.G. and Alley, M.R. Failure to demonstrate equine rhinopenumonitis virus as a cause of abortion in mares in New Zealand. NZ. Vet. J., 1976, 24:127-131.

1838. Pearson, D.A. and Herrmann, E.C. Jr. Laboratory diagnosis of rubella virus infections and antibody determinations in routine medical practice. Mayo. Clin. Proc., 1971, 46:477-483.

1839. Pecham, C.S. Clinical and serological assessment of children exposed in utero to confirmed maternal rubella. Br. Med. J., 1974, 1:259.

1840. Peck, J.E. Hearing deficit following maternal rubella. J. Pediatr., 1972, 81:414-415.

1841. Peckham, C. Cytomegalovirus infection in pregnancy. Midwife Health Visit Community Nurse, 1978, 14:297-299.

1842. Peckham, C. Cytomegalovirus: mastering a problem virus. Nurs. Mirror, 1982, 154:39-40.

1843. Peckham, C.S. Clinical and laboratory study of children exposed in utero to maternal rubella. Arch. Dis. Child, 1972, 47:571-577.

1844. Peckham, C.S. Clinical and laboratory study of children exposed in utero to maternal rubella. Arch. Dis. Child, 1972, 47:571-577.

1845. Peckham, C.S. Clinical and serological assessment of chilren exposed in utero to confirmed maternal rubella. Br. Med. J., 1974, 1:259-261.

1846. Peckham, C.S. Virus infections in pregnancy -- with special reference to rubella. Midwife Health Visit Community Nurse, 1980, 16:458-464.

1847. Peckham, C.S. Rubella and cytomegalovirus infection in pregnancy. Clin. Exp. Obstet. Gynecol., 1986, 13:97-107.

1848. Peckham, C.S., Chin, K.S., Coleman, J.C., Henderson, K., Hurley, R. and Preece, P.M. Cytomegalovirus infection in pregnancy: preliminary findings from a prospective study. Lancet, 1983, 1:1352-1355.

1849. Peckham, C. and Marshall, W.C. Rubella and other virus infections in pregnancy. J. Antimicrob. Chemother., 1979, 5:71-80.

1850. Peckham, C.S., Martin, J.A.M., Marshall, W.C. and Dudgeon, J.A. Congenital rubella deafness: a preventable disease. Lancet, 1979, 1:258-261.

1851. Peckham, C.S., Tookey, P., Nelson, D.B., Coleman, J. and Morris, N. Ethnic minority women and congenital rubella. Br. Med. J., 1983, 287:129-130.

1852. Peet, R.L., Coackley, W., Smith, V.W. and Main, C. Equine abortion associated with herpesvirus. Aust. Vet. J., 1978, 54:151.

1853. Penner, D.W. and McInnis, A.C. Intrauterine and neonatal pneumonia. Amer. J. Obst. Gyn., 1955, 69:147.

1854. Pensaert, M. and Meurichy, W. de. A porcine Enterovirus causing fetal death and mummification. II. Experimental infection of pregnant sows. Zentralbl. Veterinarmed, [B], 1973, 20:760-772.

1855. Pensaert, M., Meurichy, W. de. and Leeuwe, G. van. A porcine Enterovirus causing fetal death and mummification. I. Characteristics and identification. Zentralbl. Veterinarmed. [B], 1973, 20:749-759.

1856. Perley, N.Z. and Bills, B.J. Herpes genitalis and the childbearing cycle. M.C.N., 1983, 8:213-217.

1857. Perrin, E.C. and Gerrity, P.S. Development of children with a chronic illness. Pediatric Clinics of North America, 1984, 31:19-31.

1858. Person, D.A. The laboratory diagnosis of virus infections in gynecology and obstetrics. Clin. Obstet. Gynecol., 1972, 15:877-895.

1859. Peters, E.R. and Davis, R.L. Congenital rubella syndrome. Cerebral mineralizations and subperiosteal new bone formation as expressions of this disorder. Clin. Pediatr. (Phila.), 1966, 5:743-746.

1860. Pettay, O. Intrauterine and perinatal viral infections. A review of experiences and remaining problems. Ann. Clin. Res., 1979, 11:258-266.

1861. Pettay, O., Leinikki, P., Donner, M. and Lapinleimu, K. Herpes simplex virus infection in the newborn. Arch. Dis. Child, 1972, 47:97.

1862. Phelan, P. and Campbell, P. Pulmonary complications of rubella embryopathy. J. Pediatr., 1969, 75:202-212.

1863. Philip, A.G. and Larson, E.J. Overwhelming neonatal infection with ECHO 19 virus. J. Pediatr., 1973, 82:391-397.

1864. Phillip, J.I. Equine virus abortion. Vet. Rec., 1976, 98:283.

1865. Phillips, C.A., Maeck, J.V., Rogers, W.A. and Savel, H. Intrauterine rubella infection following immunization with rubella vaccine. J.A.M.A., 1970, 213:624-625.

1866. Phillips, C.A., Melnick, J.L., Rawls, W.E., Desmond, M.M., Yow, M.D. and Burkhardt, M. Congenital rubella syndrome: a varied and persistent infection of early life. Med. Times, 1966, 94:1411-1419.

1867. Phillips, G.A., Melnick, J.L., Yow, M.D., Bayatpour, M. and Burkhardt, M. Persistence of virus in infants with congenital rubella and in normal infants with a history of maternal rubella. J.A.M.A., 1965, 193:1027.

1868. Phuapradit, P., Varavithya, W., Theppisai, U., Chiewsilp, P., Siripunya, P. and Bunyaratavej, S. Clinical hepatitis B in infants born to HBs Ag carrier mothers. J. Med. Assoc. Thai., 1984, 67:94-97.

1869. Piccinino, F., Manzillo, G. and Sagnelli, E. The differential diagnosis between intrahepatic cholestatic jaundice and viral hepatitis during pregnancy. Acta. Hepatogastroenterol. (Stuttg.), 1975, 22:144-150.

1870. Pickard, R.E. Varicella pneumonia in pregnancy. Am. J. Obstet. Gynecol., 1968, 101:504-508.

1871. Pien, F.D., Smith, T.F., Taswell, H.F. and Hable, K.A. Cold-reactive antibodies in a case of congenital cytomegalovirus infection. Am. J. Clin. Pathol., 1974, 61:352-357.

1872. Pimentel, A.T. Rubella-caused deafness: maintaining objectivity and a positive frame of reference. Am. Ann. Deaf., 1980, 125:1012-1014.

1873. Pinching, A.J. and Jeffries, D.J. AIDS and HTLV-III/LAV infection: consequences for obstetrics and perinatal medicine. Br. J. Obstet. Gynaecol., 1985, 92:1211-1217.

1874. Pineda, R.G., Desmond, M.M., Rudolph, A.J., Halleen, W., Rawls, W. and Ziai, M.H. Impact of the 1964 rubella epidemic on a clinic population. Am. J. Obstet. Gynecol., 1968, 100:1139-1146.

1875. Piper, C.E., Ferrer, J.F., Abt, D.A. and Marshak, R.R. Postnatal and prenatal transmission of the bovine leukemia virus under natural conditions. J.N.C.I., 1979, 62:165-168.

1876. Pitt, D. and Keir, E.H. Results of rubella in pregnancy. I. Med. J. Aust., 1965, 2:647-651.

1877. Pitt, D. and Keir, E.H. Results of rubella in pregnancy. 3. Med. J. Aust., 1965, 2:737-741.

1878. Pitt, D.B. Congenital malformations and maternal rubella: progress report. Med. J. Aust., 1961, 1:881-890.

1879. Pitt, D.B. and Keir, E.H. Results of rubella in pregnancy. I. Med. J. Aust., 1965, 2:647-651, 737-741.

1880. Plager, H., Beebe, R. and Miller, J.K. Coxsackie B-5 pericarditis in pregnancy. Arch. Int. Med., 1962, 110:735.

1881. Plant, J.W., Acland, H.M. and Gard, G.P. A mucosal disease virus as a cause of abortion, hairy birth coat, and unthriftiness in sheep. 1. Infection of pregnant ewes and observations on aborted foetuses and lambs dying before one week of age. Aust. Vet. J., 1976, 52:57-63.

1882. Plant, J.W., Acland, H.M., Gard, G.P. and Walker, K.H. Clinical variations of border disease in sheep according to the source of the inoculum. Vet. Rec., 1983, 113:58-60.

1883. Plant, J.W., Byrne, D.T. and Woods, G.R. Border disease in Merino breeding flocks in western New South Wales. Aust. Vet. J., 1982, 58:78-79.

1884. Plant, J.W., Gard, G.P. and Acland, H.M. A mucosal disease virus infection of the pregnant ewe as a cause of a border disease-like condition. Aust. Vet. J., 1976, 52:247-249.

1885. Plant, J.W., Walker, K.H., Acland, H.M. and Gard, G.P. Pathology in the ovine foetus caused by an ovine pestivirus. Aust. Vet. J., 1983, 60:137-140.

1886. Plateau, E., Vannier, P. and Tillon, J.P. Atypical hog cholera infection: viral isolation and clinical study of in utero transmission. Am. J. Vet. Res., 1980, 41:2012-2015.

1887. Plotkin, S.A., Cochran, W., Lindquist, J.M., Cochran, G.G., Schaffer, D.B., Scheie, H.G. and Furukawa, T. Congenital rubella syndrome in late infancy. J.A.M.A., 1967, 200:435-441.

1888. Plotkin, S.A. and Kaye, R. Diabetes and congenital rubella. Pediatrics, 1970, 46:650-651.

1889. Plotkin, S.A., Oski, F.A., Hartnett, E.M., Hervada, A.R., Friedman, S. and Gowing, J. Some recently recognized manifestations of the rubella syndrome. J. Pediatr., 1965, 67:182-191.

1890. Plotkin, S.A. and Vaheri, A. Human fibroblasts infected with rubella virus produced a growth inhibitor. Science, 1967, 156:659.

1891. Plotz, E.J. Virus disease in pregnancy. New York J. Med., 1965, 65:1239.

1892. Plowright, W., Kalunda, M., Jessett, D.M. and Herniman, K.A. Congenital infection of cattle with the herpesvirus causing malignant catarrhal fever. Res. Vet. Sci., 1972, 13:37-45.

1893. Plummer, G. Cytomegaloviruses of man and animals. Prog. Med. Virol., 1973, 15:92.

1894. Polakoff, S. Transmission from mother to infant of hepatitis B virus infection. Midwives Chron., 1983, 96:4-5.

1895. Pollner, P. Nephrotic syndrome associated with congenital syphilis. J.A.M.A., 1966, 198:263.

1896. Pollock, R.V. and Carmichael, L.E. Maternally derived immunity to canine parvovirus infection: transfer, decline, and interference with vaccination. J. Am. Vet. Med. Assoc., 1982, 180:37-42.

1897. Pongpipat, D. Prevention of hepatitis B virus infections. Asian Pac. J. Allergy Immunol., 1986, 4:1-3.

1898. Pongpipat, D., Suvatte, V. and Assateerawatts, A. Hepatitis B immune globulin (HBIG): efficacy in the interruption of vertical transmission of hepatitis B virus (HBV) carrier state. J. Med. Assoc. Thai., 1983, 66:49-53.

1899. Pongpipat, D., Suvatte, V. and Assateerawatts, A. Vaccination against hepatitis B virus infection in neonates. Helv. Paediatr. Acta., 1984, 39:231-236.

1900. Pongpipat, D., Suvatte, V. and Assateerawatts, A. Perinatal transmission of hepatitis B virus in Thailand. Asian Pac. J. Allergy Immunol., 1985, 3:191-193.

1901. Pongpipat, D., Suvatte, V. and Assateerawatts, A. Efficacy of hepatitis-B immunoglobulin and hepatitis-B vaccine in prevention of the HBsAg carrier state in newborn infants of mothers who are chronic carriers of HBsAg and HBeAg. Asian Pac. J. Allergy Immunol., 1986, 4:33-36.

1902. Poole, C.A., Greenberg, L.A. and Mackey, E.A. The late osseous manifestations of the rubella syndrome: a case report and discussion. J. Can. Assoc. Radiol., 1966, 17:206-210.

1903. Porter, D.D., Larsen, A.E. and Porter, H.G. Reduced severity of lesions in mink infected transplacentally with Aleutian disease virus. J. Immunol., 1977, 119:872-876.

1904. Poste, G. and King, N. Isolation of a herpesvirus from the canine genital tract: association with infertility, abortion and stillbirths. Vet. Rec., 1971, 88:229-233.

1905. Potts, B.J., Osburn, B.I. and Johnson, K.P. Border disease: experimental reproduction in sheep, using a virus replicated in tissue culture. Am. J. Vet. Res., 1982, 43:1464-1466.

1906. Povey, R.C. Feline panleukopenia -- which vaccine? J. Small Anim. Pract., 1973, 14:399-406.

1907. Preblud, S.R., Kushubar, R. and Friedman, H.M. Rubella hemagglutination inhibition titers. J.A.M.A., 1982, 247:1811-1812.

1908. Preblud, S.R., Stetlen, H.C., Frank, J.A. Jr., Greaves, W.L., Hinman, A.R. and Herrmann, K.L. Fetal risk associated with rubella vaccine. J.A.M.A., 1981, 246:1413-1417.

1909. Preblud, S.R. and Williams, N.M. Fetal risk associated with rubella vaccine: implications for vaccination of susceptible women. Obstet. Gynecol., 1985, 66:121-123.

1910. Preece, M.A., Kearney, P. and Marshall, W.C. Growth hormone deficiency in the rubella syndrome. Lancet, 1977, 2:842.

1911. Preece, P.M., Blount, J.M. and Glover, J. The consequences of primary cytomegalovirus infection in pregnancy. Arch. Dis. Child, 1983, 58:970-975.

1912. Preece, P.M., Blount, J.M., Glover, J., Fletcher, G.M., Peckham, C.S. and Griffiths, P.D. The consequences of primary cytomegalovirus infection in pregnancy. Arch. Dis. Child, 1983, 58:970-975.

1913. Preece, P.M., Tookey, P. Ades, A. and Peckham, C.S. Congenital cytomegalovirus infection: predisposing maternal factors. J. Epidemiol. Community Health, 1986, 40:205-209.

1914. Price, J.M., Fisch, A.E. and Jacobson, J. Ultrasonic findings in fetal cytomegalovirus infection. J.C.U., 1978, 6:268.

1915. Prince, G.A., Potash, L., Horswood, R.L., Camargo, E., Suffin, S.C., Johnson, R.A. and Chanock, R.M. Intramuscular inoculation of live respiratory syncytial virus induces immunity in cotton rats. Infect. Immun., 1979, 23:723-728.

1916. Pritchard, G.C. Observations on clinical aspects of transmissible gastroenteritis of pigs in Norfolk and Suffolk, 1980-81. Vet. Rec., 1982, 110:465-469.

1917. Prober, C.G., Sullender, W.M., Yasukawa, L.L., Au, D.S., Yeager, A.S. and Arvin, A.M. Low risk of herpes simplex virus infections in neonates exposed to the virus at the time of vaginal delivery to mothers with recurrent genital herpes simplex virus infections. N. Engl. J. Med., 1987, 316:240-244.

1918. Prout, B.J. and Marks, V. Infective hepatitis and acquired hypogammaglobulinaemia during pregnancy with complete recovery. Postgrad. Med. J., 1967, 43:492-495.

1919. Prozesky, L., Thomson, G.R., Gainaru, M.D., Herr, S. and Kritzinger, L.J. Lesions resulting from inoculation of porcine foetuses with porcine parvovirus. Onderstepoort. J. Vet. Res., 1980, 47:269-274.

1920. Prozesky, O.W., Szmuness, W., Stevens, C.E., Kew, M.C., Harley, E.J., Hoyland, J.A., Scholtz, J.E., Mitchell, A.D., Shabangu, A., Kunene, E. et al. Baseline epidemiological studies for a hepatitis B vaccine trial in Kangwane. S. Afr. Med. J., 1983, 64:891-893.

1921. Prydie, J. Resistance against canine distemper induced by measles virus. I. Measles vaccination in the presence of distemper antibodies. Vet. Rec., 1968, 83:554-559.

1922. Puck, J.M., Glezen, W.P., Frank, A.L. and Six, H.R. Protection of infants from infection with influenza A virus by transplacentally acquired antibody. J. Infect. Dis., 1980, 142:844-849.

1923. Purtilo, D.T., Bhawan, J., Liao, S., Brutus, A., Yang, J.P. and Balogh, K. Fatal varicella in a pregnant woman and a baby. Am. J. Obstet. Gynecol., 1977, 127:208-209.

1924. Purtilo, D.T. and Sakamoto, K. Reactivation of Epstein-Barr virus in pregnant women: social factors, and immune competence as determinants of lymphoproliferative diseases - a hypothesis. Med. Hypotheses, 1982, 8:401-408.

1925. Purvis-Smith, S.G., Hayes, K. and Menser, M.A. Dermatoglyphics in children with prenatal cytomegalovirus infection. Lancet, 1972, 2:976-977.

1926. Purvis-Smith, S.G. and Menser, M.A. Dermatoglyphics in adults with congenital rubella. Lancet, 1968, 2:141-143.

1927. Puvacic, Z., Gaon, J., Bryan, J. and Kendrick, M. Study of causes of high case fatality and of outcome of pregnancy in viral hepatitis. Acta. Med. Lugosl., 1977, 31:305-320.

# Q

1928. Quaggin, A. Get prepared for more cases of AIDS during pregnancy. Can. Med. Assoc. J., 1987, 136:192-193.

1929. Quinn, P.T. and Lofberg, J.V. Maternal herpetic breast infection: another hazard of neonatal herpes simplex. Med. J. Aust., 1978, 2:411-412.

1930. Quinnan, G.V. Jr. and Kelman, A.D. Rubella vaccination during pregnancy. N. Engl. J. Med., 1976, 295:396.

# R


1931. Rabinowitz, J.G., Wolf, B.S., Greenberg, E.I. and Rausen, A.R. Osseous changes in rubella embryopathy. Radiology, 1965, 85:494-500.

1932. Rachelefsky, G.S. and Herrmann, K.L. Congenital rubella surveillance following epidemic rubella in a partially vaccinated community. J. Pediatr., 1974, 84:474-478.

1933. Rajhvajn, B., Krznar, B., Stoiljkovic, C., Orescanin, M. and Smerdel, S. Vaccination against smallpox in early pregnancy. Acta. Med. Lugosl., 1973, 27:351-357.

1934. Raker, R.K., Steinberg, S., Drusin, L.M. and Gershon, A. Antibody to varicella zoster virus in low-birth-weight newborn infants. J. Pediatr., 1978, 93:505-506.

1935. Ramanathan, S., Sheth, R. and Turndorf, H. Anesthesia for cesarean section in patients with genital herpes infections: a retrospective study. Anesthesiology, 1986, 64:807-809.

1936. Rambar, A.C. Effect of maternal virus infections on the fetus. II. III Med. J., 1969, 136:599-605.

1937. Rambar, A.C. Effect of maternal virus infections on the fetus. III Med. J., 1969, 136:261-267.

1938. Ramia, S., Abdul-Jabbar, F., Bakir, T.M. and Hossain, A. Vertical transmission of hepatitis B surface antigen in Saudi Arabia. Ann. Trop. Paediatr., 1984, 4:213-216.

1939. Ramphal, R., Donnelly, W.H. and Small, P.A. Fatal influenzal pneumonia in pregnancy: failure to demonstrate transplacental transmission of influenza virus. Am. J. Obstet. Gynecol., 1980, 138:347-348.

1940. Ramsay, A.M. Rubella syndrome. Postgrad. Med. J., 1966, 42:83-85.

1941. Randolph, V.L. and Heath, C.W. Jr. Influenza during pregnancy in relation to subsequent childhood leukemia and lymphoma. Am. J. Epidemiol., 1974, 100:399-409.

1942. Rao, A.R., Sukumar-Savithri, M.S., Mascreen, R.K., Kamalakshi, S., Paramasivam, T.V., Parasuraman, A.R. and Shantha, M. Vaccination and pregnancy: placental transmission of antibodies. Indian J. Med. Res., 1969, 57:1250-1260.

1943. Rao, A.V., Devi, C.S., Savithri, P. and Seshirekha, E. Infectious hepatitis in pregnancy and puerperium -- a study of 60 cases. Indian J. Med. Sci., 1969, 23:471-478.

1944. Rauth, J.L., Schiff, G.M. and Johnson, L.B. Rubella surveillance and immunization among adolescent girls in Cincinnati. Am. J. Dis. Child, 1972, 124:71-75.

1945. Ravindran, R.S., Gupta, C.D. and Stoops, C.A. Epidural analgesia in the presence of herpes simplex virus (type 2) infection. Anesth. Analg., 1982, 61:714-715.

1946. Rawls, W.E. Congenital rubella: the significance of virus persistence. Prog. Med. Virol., 1968, 10:238-285.

1947. Rawls, W.E., Desmyter, J. and Melnick, J.L. Serologic diagnosis and fetal involvement in maternal rubella. Criteria for abortion. J.A.M.A., 1968, 203:627-631.

1948. Rawls, W.E., Desmyter, J. and Melnick, J.L. Virus carrier cells and virus-free cells in fetal rubella. Proc. Soc. Exp. Biol. Med., 1968, 129:477-483.

1949. Rawls, W.E. and Melnick, J.L. Rubella virus carrier cultures derived from congenitally infected infants. J. Exp. Med., 1966, 123:795-816.

1950. Rawls, W.E. and Melnick, J.L. Intrauterine rubella transmission. J.A.M.A., 1968, 204:1009-1010.

1951. Rawls, W.E., Phillips, A., Melnick, J.L. and Desmond, M.M. Persistent virus infection in congenital rubella. Arch. Ophthalmol., 1967, 77:430-433.

1952. Rawls, W.E., Wong, C.L., Blajchman, M., Venturelli, J., Watts, J., Chernesky, M. and Saigal, S. Neonatal cytomegalovirus infections: the relative role of neonatal blood transfusion and maternal exposure. Clin. Invest. Med., 1984, 7:13-19.

1953. Ray, D.A., Evans, A.T., Elliott, J.P. and Freeman, R.K. Maternal herpes infection complicated by prolonged premature rupture of membranes. Am. J. Perinatol., 1985, 2:96-100.

1954. Ray, G.C. and Wedgewood, R.J. Neonatal listeriosis. Six case reports and a review of the literature. Pediatrics, 1964, 34:378.

1955. Ray, S. Transplacental transmission of polio antibodies -- flocculating and neutralizing. Indian J. Exp. Biol., 1969, 7:98-102.

1956. Rayman, G., Spencer, P., Russell, M.E. and Ward, P.D. Finger sepsis. Br. Med. J. [Clin. Res.], 1983, 287:215.

1957. Raymond, F.R. and Thomas, R.J. Condyloma acuminatum in pregnancy. J. Am. Osteopath. Assoc., 1983, 82:774-778.

1958. Redeker, A.G. Viral hepatitis: current concepts. Postgrad. Med., 1973, 53:77-81.

1959. Redman, D.R., Bohl, E.H. and Cross, R.F. Intrafetal inoculation of swine with transmissible gastroenteritis virus. Am. J. Vet. Res., 1978, 39:907-911.

1960. Redman, D.R., Bohl, E.H. and Ferguson, L.C. Porcine parvovirus: natural and experimental infections of the porcine fetus and prevalence in mature swine. Infect. Immun., 1974, 10:718-723.

1961. Reece, R. An antibody test for rubella. Minn. Med., 1971, 54:318.

1962. Reed, D.E., Bicknell, E.J., Larson, C.A., Knudtson, W.U. and Kirkbride, C.A. Infectious bovine rhinotracheitis virus-induced abortion: rapid diagnosis by fluorescent antibody technique. Am. J. Vet. Res., 1971, 32:1423-1426.

1963. Reed, D.E., Langpap, T.J. and Bergeland, M.E. Bovine abortion associated with mixed Movar 33/63 type herpesvirus and bovine viral diarrhea virus infection. Cornell Vet., 1979, 69:54-66.

1964. Rees, R.B. Warts -- a clinical's view. Cutis, 1981, 28:175-180.

1965. Reesink, H.W., Reerink-Brongers, E.E., Lafeber-Schut, B.J., Kalshoven-Benschop, J. and Brummelhuis, H.G. Prevention of chronic HBsAg carrier state in infants of HBsAg-positive mothers by hepatitis B immunoglobulin. Lancet, 1979, 2:436-438.

1966. Reeves, A.M. Rubella hemagglutination inhibition testing program. J. Iowa Med. Soc., 1971, 61:491-493.

1967. Reid, D. The prophylactic value of gamma globulin in rubella. Public Health, 1967, 81:280-284.

1968. Reid, W.M. Congenital rubella -- on approach to prevention. Can. Nurse, 1971, 67:38-40.

1969. Reikvam, T.M., Rud, T. and Skaug, V. Intra-uterine herpes simplex virus infection. Premature newborn with disseminated infection at delivery. Zentralbl. Gynakol., 1981, 103:1473-1475.

1970. Reisinger, K.S., Evans, P., Yost, G. and Rogers, K.D. Congenital tuberculosis: Report of a case. Pediatrics, 1974, 54:74.

1971. Rekant, S.I. Eczema herpeticum and pregnancy. Obstet. Gynecol., 1973, 41:387-391.

1972. Reman, O., Freymuth, F., Laloum, D. and Bonte, J.F. Neonatal respiratory distress due to mumps. Arch. Dis. Child, 1986, 61:80-81.

1973. Remington, J.S., Jacobs, L. and Melton, M.L. Congenital transmission of toxoplasmosis from mother animals with acute and chronic infections. J. Infect. Dis., 1961, 108:163-173.

1974. Remington, J.S., Miller, M.J. and Brownlee, I. IgM antibodies in acute toxoplasmosis. I. Diagnostic significance in congenital cases and a method for their rapid demonstration. Pediatrics, 1968, 41:1082-1091.

1975. Renis, H.E. Genital infection of female hamsters with herpesvirus hominis type 2. Proc. Soc. Exp. Biol. Med., 1975, 150:723.

1976. Renne, R.A., McLaughlin, R. and Jenson, A.B. Measles virus-associated endometritis, cervicitis, and abortion in a rhesus monkey. J. Am. Vet. Med. Assoc., 1973, 163:639-641.

1977. Reyes, M.P., Ostrea, E.M. Jr., Roskamp, J. and Lerner, A.M. Disseminated neonatal ECHOvirus 11 disease following antenatal maternal infection with a virus-positive cervix and virus-negative gastrointestinal tract. J. Med. Virol., 1983, 12:155-159.

1978. Reyes, M.P., Zalenski, D., Smith, F., Wilson, F.M. and Lerner, A.M. Coxsackievirus-positive cervices in women with febrile illnesses during the third trimester in pregnancy. Am. J. Obstet. Gynecol., 1986, 155:159-161.

1979. Reynolds, D.W., Stagno, S., Hosty, T.S., Tiller, M. and Alford, C.A. Jr. Maternal cytomegalovirus excretion and perinatal infection. N. Engl. J. Med., 1973, 289:1-5.

1980. Reynolds, D.W., Stagno, S., Reynolds, R. and Alford, C.A. Jr. Perinatal cytomegalovirus infection: influence of placentally transferred maternal antibody. J. Infect. Dis., 1978, 137:564-567.

1981. Reynolds, D.W., Stagno, S. and Stubbs, K.G. Inapparent congenital cytomegalovirus infection with elevated cord IgM levels: causal relations with auditory and mental deficiency. N. Engl. J. Med., 1974, 290:291-296.

1982. Reynolds, D.W., Stagno, S., Stubbs, K.G., Dahle, A.J., Livingston, M.H., Saxon, S.S. and Alford, C.A. Inapparent congenital cytomegalovirus infection with elevated cord IgM levels. N. Engl. J. Med., 1974, 290:291.

1983. Reynolds, D.W., Stagno, S., Stubbs, K.G., Dahle, A.J., Livingston, M.M., Saxon, S.S. and Alford, C.A. Congenital cytomegalovirus infection: relation to auditory and mental deficiency. N. Engl. J. Med., 1974, 290:291-296.

1984. Reynolds, D.W., Stark, E.W. and Borton, T.E. Hearing loss after maternal rubella: need for accurate diagnosis of maternal infection. Pediatrics, 1970, 46:159-160.

1985. Richardson, C., Hebert, C.N. and Done, J.T. Experimental Border disease in sheep: dose-response effect. Br. Vet. J., 1976, 132:202-208.

1986. Richardson, C., Taylor, W.P., Terlecki, S. and Gibbs, E.P. Observations on transplacental infection with bluetongue virus in sheep. Am. J. Vet. Res., 1985, 46:1912-1922.

1987. Richardson, J. Suspected transplacental Coxsackie infection. Lancet, 1977, 2:877-878.

1988. Richter-Reichhelm, H.B., Trautwein, G., VonBenten, K., Liess, B. and Frey, H.R. Experimental transplacental transmission of hog cholera virus in pigs. II. Immunopathological findings in the fetus. Zentralbl. Veterinarmed. [B], 1980, 27:243-252.

1989. Rinaldo, C.R. Jr., Isackson, D.W., Overall, J.C. Jr., Glasgow, L.A., Brown, T.T., Bistner, S.I., Gillespie, J.H. and Scott, F.W. Fetal and adult bovine interferon production during bovine viral diarrhea virus infection. Infect. Immun., 1976, 14:660-666.

1990. Ringe, M.E. Poliomyelitis in pregnancy: a report of 79 cases in Connecticut. N. Engl. J. Med., 1957, 256:281.

1991. Rinvik, R. Congenital varicella encephalomyelitis in surviving newborn. Am. J. Dis. Child, 1969, 117:231-235.

1992. Ritchie, A.E., Bryner, J.H. and Foley, J.W. Role of DNA and bacteriophage in Campylobacter auto-agglutination. J. Med. Microbiol., 1983, 16:333-340.

1993. Rivera, E., Sjoland, L. and Karlsson, K.A. A solid phase fluorescent immunoassay for the rapid detection of virus antigen or antibodies in fetuses infected with porcine parvovirus. Arch. Virol., 1986, 88:19-26.

1994. Robb, J.A., Benirschke, K. and Barmeyer, R. Intrauterine latent herpes simplex virus infection. I. Spontaneous abortion. Hum. Pathol., 1986, 17:1196-1209.

1995. Robb, J.A., Benirschke, K., Mannino, F. and Voland, J. Intrauterine latent herpes simplex virus infection: II. Latent neonatal infection. Hum. Pathol., 1986, 17:1210-1217.

1996. Roberts, J.K. A case of genital herpesvirus infection in pregnancy. J. Obstet. Gynaecol. Br. Commonw., 1973, 80:188-190.

1997. Roberts, W., Grist, N.R. and Giroud, P. Human abortion associated with infection by ovine abortion agent. Br. Med. J., 1967, 4:37.

1998. Roberts, W.E. and Morrison, J.C. Viral hepatitis B in pregnancy: a case report. J. Miss. State Med. Assoc., 1985, 26:93-94.

1999. Robertson, G.G., Williamson, A.P. and Blattner, R.J. Origin and development of lens cataracts in mumps-infected chick embryos. Amer. J. Anat., 1964, 115:473.

2000. Robertson, J.R. and Bucknall, A.B.V. Pregnancy and HTLV-III-LAV transmission in heroin users. Health Bull. (EDINB), 1987, 44:364-366.

2001. Robertson, J.S. and Sheard, A.V. Altered sex ratio after an outbreak of hepatitis. Lancet, 1973, 1:432-534.

2002. Robertson, L.D. Management of herpes during pregnancy. Health Care Women Int., 1987, 8:55-63.

2003. Roche, T.F., Sheehan, P., Lydia, M., Walsh, J. and MacAirt, J. A study of handicaps and their effect on lip reading among rubella deaf girls. A preliminary report. Dev. Med. Child Neurol., 1971, 13:497-507.

2004. Rodin, A.E. and McGanity, W.J. Clinicopathologic conference. Tex Med., 1967, 63:62-66.

2005. Rogers, M.F., Ewing, E.P., Jr., Warfield, D., Hardy, A.M., Emery, D.R. and Wolf, G.C. Virologic studies of HTLV-III/LAV in pregnancy: case report of a woman with AIDS. Obstet. Gynecol., 1986, 68:2S-6S.

2006. Rogers, S.C. Influenza and congenital abnormalities. Lancet, 1972, 1:261.

2007. Roman-Campos, G., Navarro-de-Roman, L.I., Toro, G. and Vergara, I. Herpes encephalitis in pregnancy. Am. J. Obstet. Gynecol., 1979, 135:158-159.

2008. Romano, A., Weinberg, M., Bar-Izhak, R., Mashiah, S. and Eylan, E. Rate and various aspects of eye infection resulting from congenital rubella. J. Pediatr. Ophthalmol. Strabismus., 1979, 16:26-30.

2009. Romvary, J., Toth, B. and Vizy, L. Isolation and properties of equine rhinopneumonitis virus. Acta. Vet. Acad. Sci. Hung., 1969, 19:311-317.

2010. Rones, B. The relationship of German measles during pregnancy to congenital ocular defects. Med. Ann. D.C., 1944, 13:285.

2011. Rorke, L.B., Fabiyi, A., Elizan, T.S. and Sever, J.L. Experimental cerebrovascular lesions in congenital and neonatal rubella-virus infections of ferrets. Lancet, 1968, 2:153-154.

2012. Rorke, L.B. and Spiro, A.J. Cerebral lesions in congenital rubella syndrome. J. Pediatr., 1967, 70:243-255.

2013. Rosen, E.U., Abrahams, C. and Rabinowitz, L. Nephropathy of congenital syphilis. S. Afr. Med. J., 1973, 47:1606.

2014. Rosenberg, H.S., Kohl, S. and Vogler, C. Viral infections of the fetus and the neonate. Monogr. Pathol., 1981, 133-200.

2015. Rosenberg, H.S., Oppenheimer, E.H. and Esterly, J.R. Congenital rubella syndrome: the late effects and their relation to early lesions. Perspect. Pediatr. Pathol., 1981, 183-202.

2016. Rosendahl, C., Kochen, M.M, Kretschmer, R., Wegscheider, K. and Kaiser, D. Avoidance of perinatal transmission of hepatitis B virus: is passive immunisation always necessary? Lancet, 1983, 1:1127-1129.

2017. Rosenthal, D. and Morris, N. Varicella, herpes zoster infections, and congenital defects. Arch. Dermatol., 1973, 108:427.

2018. Rosenthal, P. Mothers and hepatitis B. Pediatrics, 1983, 72:265-266.

2019. Rosner, S.F. Infectious bovine rhinotracheitis: clinical review, immunity, and control. J. Am. Vet. Med. Assoc., 1968, 153:1631-1638.

2020. Ross, C.A. Cardiomyopathy of pregnancy. Br. Med. J., 1968, 4:453-454.

2021. Ross, C.A. Rubella and pregnancy. Br. Med. J., 1968, 2:432.

2022. Ross, C.A., Bell, E.J., Kerr, M.M. and Williams, K.A. Infective agents and embryopathy in the west of Scotland 1966-1970. Scott Med. J., 1972, 17:252-258.

2023. Ross, C.E., Dubovi, E.J. and Donis, R.O. Herd problem of abortions and malformed calves attributed to bovine viral diarrhea. J. Am. Vet. Med. Assoc., 1986, 188:618-619.

2024. Ross, L. Incidental finding of cytomegalovirus inclusions in cervical glands. A case report. Am. J. Obstet. Gynecol., 1966, 95:956-958.

2025. Ross, L.J. Fingerprints in congenital rubella following maternal gamma globulin. Acta. Paediatr. Scand., 1979, 68:71-74.

2026. Rossi, C.R., Kiesel, G.K. and Hubbert, W.T. Immunoglobulin concentrations and bovine enterovirus inhibitors in fetal bovine fluids. Cornell Vet., 1976, 66:381-386.

2027. Rossier, E., Phipps, P.H. and Weber, J.M. Persistence of naturally acquired cell-mediated immunity to rubella virus during the first trimester of pregnancy. Can. J. Microbiol., 1983, 29:1680-1682.

2028. Rotthauwe, H.W. Viral hepatitis in infancy and childhood. Clin. Gastroenterol., 1974, 3:437-452.

2029. Rowan, D.F., McCraw, M.F. and Eward, R.D. Virus infections during pregnancy. Obstet. Gynecol., 1968, 32:356-364.

2030. Rowe, R.D. Cardiovascular disease in the rubella syndrome. Cardiovasc. Clin., 1973, 5:61-80.

2031. Rowe, W.P., Hartley, J.W., Waterman, S., Turner, H.C. and Huebner, R.J. Cytopathogenic agent resembling human salivary gland virus recovered from tissue cultures of human adenoids. Proc. Soc. Exp. Biol. Med., 1956, 92:418.

2032. Rowen, M., Singer, M.I. and Moran, E.T. Intracranial calcification in the congenital rubella syndrome. Amer. J. Roentgenol., 1972, 115:86.

2033. Roy, F.H., Fuste, F., Hiatt, R.L., Deutsch, A.R. and Korones, S.B. The congenital rubella syndrome with virus recovery. Ocular pathology and literature review. Am. J. Ophthalmol., 1966, 62:222-232.

2034. Ruben, F.L. and Thompson, D.S. Cord blood lymphocyte in vitro responses to influenza A antigens after an epidemic of influenza A/Port Chalmers/73 (H3N2). Am. J. Obstet. Gynecol., 1981, 141:443-447.

2035. Ruben, R.J. Screening for rubella during pregnancy. N. Engl. J. Med., 1970, 283:1292.

2036. Rubinstein, A., Sicklick, M., Gupta, A., Bernstein, L., Klein, N., Rubinstein, E., Spigland, I., Fruchter, L., Litman, N., Lee, H. and Hollander, M. Acquired immunodeficiency with reversed T4/T8 ratios in infants born to promiscuous and drug-addicted mothers. Journal of the American Medical Association, 1983, 249:2350-2356.

2037. Rubinstein, A., Sicklick, M., Gupta, A., Bernstein, L., Klein, N., Rubinstein, E., Spigland, I., Fruchter, L., Litman, N., Lee, H. and Hollander, M. Acquired immunodeficiency syndrome in infants. American Journal of Diseases of Childhood, 1983, 137:825-827.

2038. Rudolph, A.J. and Desmond, M.M. Clinical manifestations of the congenital rubella syndrome. Int. Ophthalmol. Clin., 1972, 12:3-19.

2039. Rudolph, A.J., Yow, M.D., Phillips, A., Desmond, M.M., Blattner, R.J. and Melnick, J.L. Transplacental rubella infection in newly born infants. J.A.M.A., 1965, 191:843.

2040. Rushton, D.I., Collie, M.H., Sweet, C., Husseini, R.H. and Smith, H. The effects of maternal influenzal viraemia in late gestation on the conceptus of the pregnant ferret. J. Pathol., 1983, 140:181-191.

2041. Rutter, D., Griffiths, P. and Trompeter, R.S. Cytomegalic inclusion disease after recurrent maternal infection. Lancet, 1985, 2:1182.

2042. Ryan, M.J., OConnor, D.J. and Nielsen, S.W. Staphylococcus aureus mastitis in nursing mink affected with aleutian disease. J. Wildl. Dis., 1979, 15:533-535.

# S

2043. Sabau, M., Capilna, E., Demeter, S. and Szilagyi, I. Incidence of HBeAg in asymptomatic carrier mothers and vertical transmission of HBV to their infants. Rev. Med. Chir. Soc. Med. Nat. Iasi., 1979, 83:259-262.

2044. Sabau, M., Capilna, Kiss, E. Prevalence of HBsAg and anti-HBs among delivering women. Virologie, 1979, 30:121-124.

2045. Sabin, A.B., Flores-Arechiga, A., Fernandez-de-Castro, J., Sever, J.L., Madden, D.L., Shekarchi, I. and Albrecht, P. Successful immunization of children with and without maternal antibody by aerosolized measles vaccine. I. Different results with undiluted human diploid cell and chick embryo fibroblast vaccines. J.A.M.A., 1983, 249:2651-2662.

2046. Sabine, M., Feilen, C., Herbert, L., Jones, R.F., Lomas, S.W., Love, D.N. and Wild, J. Equine herpesvirus abortion in Australia 1977 to 1982. Equine Vet. J., 1983, 15:366-370.

2047. Sachdeva, R. Congenital rubella syndrome at Mombasa. East Afr. Med. J., 1973, 50:146-151.

2048. Sacher, M., Eder, G., Baumgarten, K. and Thaler, H. Infectivity of anti-hepatitis Be-positive hepatitis B carrier mothers. Pediatrics, 1983, 72:266-267.

2049. Sacker, I., Walker, M. and Brunell, P.A. Abscess in newborn infant caused by mycoplasma. Pediatrics, 1970, 46:303.

2050. Sacks, J.J., Olson, B., Soter, J. and Clark, C. Employee rubella screening programs in Arizona hospitals. J.A.M.A., 1983, 249:2675-2678.

2051. Sadoff, L. Antimalarial drugs and Burkitt's lymphoma. Lancet, 1973, 2:1262-1263.

2052. Saif, L.F. and Bohl, E.H. Immunoglobulin classes of antibodies in milk of swine after intranasal exposure to pseudorabies virus or transmissible gastroenteritis virus. Infect. Immun., 1977, 16:961-966.

2053. Saigal, S., Lunyk, O., Larke, R.P.B. The outcome in children with congenital cytomegalovirus infection. Am. J. Dis. Child, 1982, 136:896-901.

2054. Saigal, S., Lunyk, O., Larke, R.P.B. and Chernesky, M.A. The outcome of children with congenital cytomegalovirus infection: a longitudinal follow-up study. Am. J. Dis. Child, 1982, 136:896-901.

2055. Saint-Yves, I.F., Crouch, E. and Rumold, M. Sequelae following smallpox vaccination in the Sepik District of the Territory of Papua and New Guinea. Med. J. Aust., 1967, 1:599-601.

2056. Saito, K., Koizumi, F. and Sumiyoshi, Y. Viral placentitis - a case report. Acta. Pathol. Jpn., 1977, 27:275-282.

2057. Sak, R.J. and Ruben, R.J. Effects of recurrent middle ear effusion in preschool years on language and learning. Journal of Developmental and Behavioral Pediatrics, 1982, 3:7-11.

2058. Salerno, R.A., Larson, V.M., Phelps, A.H. and Hilleman, M.R. Infection and immunization of cats with the Kawakami-Theilen strain of feline leukemia virus. Proc. Soc. Exp. Biol. Med., 1979, 160:18-23.

2059. Sallomi, S.J. Rubella in pregnancy. A review of prospective studies from the literature. Obstet. Gynecol., 1966, 27:252-256.

2060. Salvioli, G.P., Faldella, G., Alessandroni, R. and Larari, M. Prevention of perinatal transmission of chronic hepatitis B surface antigen (HBsAg) carrier state. Pediatrics, 1984, 73:408-410.

2061. Sander, J.E. and Chaplin, E.R. Porencephaly and cytomegalovirus in a surviving monozygotic twin. J. Pediatr., 1981, 99:500.

2062. Sandow, D., Okubagzhi, G.S., Arnold, U. and Denkmann, N. Seroepidemiological study in rubella in pregnant women in Gondar Region, northern Ethiopia. Ethiop. Med. J., 1982, 20:173-178.

2063. Sangiorgio, P. and Weissenbacher, M.C. Congenital and perinatal infection with Junin virus in guinea pigs. J. Med. Virol., 1983, 11:161-165.

2064. Sansom, B.F., Symonds, H.W. and Taylor, P.J. The transfer of copper from ewes with Border disease to their lambs. J. Comp. Pathol., 1979, 89:361-366.

2065. Sarak, J.K., Chakravarty, S.K., Ray, S.N., Ray, I., Chakravarty, M.S., Mukherjee, K.K., Mitra, A.C. and Mukherjee, M.K. Transplant transmission of haemagglutination-inhibiting antibodies of pox and arboviruses. Indian J. Med. Res., 1973, 61:1153-1157.

2066. Sarkar, J.K., Mukherjee, K.K., Chakravarty, M.S., Roy, I., Bose, A.K., Chaudhuri, A., Das, B.C. and Chakravarty, S.K. Virologically proved rubella cases in Calcutta. J. Indian Med. Assoc., 1974, 63:194-196.

2067. Sato, H., Albrecht, P., Reynolds, D.W., Stagno, S. and Ennis, F.A. Transfer of measles, mumps, and rubella antibodies from mother to infant. Its effect on measles, mumps, and rubella immunization. Am. J. Dis. Child, 1979, 133:1240-1243.

2068. Sato, H., Hishiyama, M. and Shishido, A. Growth of rubella, measles and mumps viruses in rat fetus. Jpn. J. Med. Sci. Biol., 1976, 29:39-44.

2069. Sato, K., Inaba, Y., Tokuhisa, S., Miura, Y., Akashi, H. and Tanaka, Y. Antibodies against several viruses in sera from normal bovine fetuses and precolostral calves. Natl. Inst. Anim. Health Q. (Tokyo), 1980, 20:77-78.

2070. Sattar, S.A., Bohl, E.H. and Senturk, M. Viral causes of bovine abortion in Ohio. J. Am. Vet. Med. Assoc., 1965, 147:1207-1210.

2071. Sattar, S.A., Bohl, E.H. and Trapp, A.L. Abortion in cattle caused by experimental infection with infectious bovine rhinotracheitis virus. Cornell Vet., 1967, 57:438-454.

2072. Savage, M.O., Moosa, A. and Gordon, R.R. Maternal varicella infection as a cause of fetal malformations. Lancet, 1973, 1:352-354.

2073. Saxen, I. The association between maternal influenza, drug consumption and oral clefts. Acta. Odontol. Scand., 1975, 33:259-267.

2074. Saxen, L., Hjelt, L., Sjostedt, J.E., Hakosalo, J. and Hakosalo, J. Asian influenza during pregnancy and congenital malformations. Acta Path. Microbiol. Scand., 1960, 49:114.

2075. Saxon, S.A., Knight, W., Reynolds, D.W., Stagno, S. and Alford, C.A. Intellectual deficits in children born with subclinical congenital toxoplasmosis: a preliminary report. J. Pediatr., 1973, 82:792-797.

2076. Saxoni, F., Lapatsanis, P. and Pantelakis, S.N. Congenital syphilis: a description of 18 cases and re-examination of an old but ever-present disease. Clin. Pediat., 1967, 6:687.

2077. Scarselli, G., Gargiulo, A., Branconi, F. and Di-Tommaso, M. Post conception failures in reproduction: infectious diseases and immunitary problems. Acta. Eur. Fertil., 1984, 15:363-367.

2078. Schaefer, J. and Shaw, E.B. Poliomyelitis in pregnancy. Calif. Med., 1949, 70:16.

2079. Schaeffer, M., Fox, M.J. and Li, C.P. Intrauterine poliomyelitis infection. J.A.M.A., 1954, 155:248.

2080. Scheie, H.G., Schaffer, D.B., Plotkin, S.A. and Kertesz, E.D. Congenital rubella cataracts. Arch. Ophthalmol., 1967, 77:440-444.

2081. Scher, J., Bottone, E., Desmond, E. and Simons, W. The incidence and outcome of asymptomatic herpes simplex genitalis in an obstetric population. Am. J. Obstet. Gynecol., 1982, 144:906-909.

2082. Schidlovsky, G. and Ahmed, M. C-type virus particles in placentas and fetal tissues of Rhesus monkeys. J. Natl. Cancer Inst., 1973, 51:225-233.

2083. Schiefer, B. Bovine abortion associated with Cytomegalovirus infection. Zentralbl. Veterinarmed. [B], 1974, 21:145-151.

2084. Schiff, E.R. Epidemiology of virus B hepatitis. Med. Clin. North Am., 1975, 59:835-842.

2085. Schiff, G.M. and Dine, M.S. Transmission of rubella from newborns. Am. J. Dis. Child, 1965, 110:447-451.

2086. Schiff, G.M., Sutherland, J. and Light, I. Congenital rubella. In: Thalhammer, O. (ed.): Prenatal infections. International Symposium of Vienna, September 2-3, 1970, Stuttgart, Georg Thieme Verlag, 1971, pp. 31-36.

2087. Schiff, G.M., Sutherland, J.M., Light, I.J. and Bloom, J.E. Studies on congenital rubella. Preliminary results on the frequency and significance of presence of rubella virus in the newborn and the effect of gamma-globulin in preventing congenital rubella. Am. J. Dis. Child, 1965, 110:441-443.

2088. Schlafer, D.H., Schultz, R.D., Scott, F.W. and Duncan, J.R. Bovine fetal inoculations with calf rotavirus. Can. J. Comp. Med., 1979, 43:405-414.

2089. Schmidt, R.E. and Panciera, R.J. Cerebral malformation in fetal lambs from a bluetongue-enzootic flock. J. Am. Vet. Med. Assoc., 1973, 162:567-568.

2090. Schneider, J. Virus disease in pregnancy. Nebr. Med. J., 1969, 54:98-102.

2091. Schoenbaum, S.C. Problems of rubella control. N. Engl. J. Med., 1970, 283:819.

2092. Schoenbaum, S.C., Biano, S. and Mack, T. Epidemiology of congenital rubella syndrome. The role of maternal parity. J.A.M.A., 1975, 233:151-155.

2093. Schoenbaum, S.C. and Weinstein, L. Respiratory infection in pregnancy. Clin. Obstet. Gynecol., 1979, 22:293-300.

2094. Schoenfeld, A., Ovadia, L., Levavi, C., Nitke, S., Wallach, D., Schattner, A. and Revel, M. Identification of viral infections in pregnancy by assay of (2'-5')-oligo-isoadenylate synthetase. Early Hum. Dev., 1985, 10:279-286.

2095. Schopfer, K., Lauber, E. and Krech, U. Congenital cytomegalovirus infection in newborn infants of mothers infected before pregnancy. Arch. Dis. Child, 1978, 53:536-539.

2096. Schoub, B.D., Prozesky, O.W., Lecatsas, G. and Oosthuizen, R. The role of breast-feeding in the prevention of rotavirus infection. J. Med. Microbiol., 1978, 11:25-31.

2097. Schwartz, H.A. Mumps in pregnancy. Am. J. Obstet. Gynecol., 1950, 60:875.

2098. Schwartz, R.W. Clinical considerations of rubella in pregnancy. NY. State J. Med., 1968, 68:392-397.

2099. Schweitzer, I.L. Vertical transmission of the hepatitis B surface antigen. Am. J. Med. Sci., 1975, 270:287-291.

2100. Schweitzer, I.L., Dunn, A.E., Peters, R.L. and Spears, R.L. Viral hepatitis b in neonates and infants. Am. J. Med., 1973, 55:762-771.

2101. Schweitzer, I.L., Edwards, V.M. and Brezina, M. e Antigen in HBs Ag-carrier mothers. N. Engl. J. Med., 1975, 293:940.

2102. Schweitzer, I.L., Mosley, J.W., Ashcaval, M., Edwards, V.M. and Overby, L.B. Factors influencing neonatal infection by hepatitis B virus. Gastroenterology, 1973, 65:277-283.

2103. Schweitzer, I.L. and Peters, R.L. Pregnancy in hepatitis B antigen positive cirrhosis. Obstet. Gynecol., 1976, 48:53S-56S.

2104. Schweitzer, I.L. and Spears, R.L. Hepatitis-associated antigen (Australia antigen) in mother and infant. N. Engl. J. Med., 1970, 283:570-572.

2105. Schweitzer, I.L., Wing, A., McPeak, C. and Spears, R.L. Hepatitis and hepatitis-associated antigen in 56 mother-infant pairs. J.A.M.A., 1972, 220:1092-1095.

2106. Scott, F.M. and Holliman, A. Serum antibodies to bovine mammillitis virus in pregnant heifers. Vet. Rec., 1984, 114:19.

2107. Scott, F.W. Comments on feline panleukopenia biologics. J. Am. Vet. Med. Assoc., 1971, 158:910-915.

2108. Scott, F.W. and Gillespie, J.H. Immunization for feline panleukopenia. Vet. Clin. N. Am., 1971, 1:231-240.

2109. Scott, F.W., Kahrs, R.F., DeLahunte, A., Brown, T.T., McEntee, K. and Gillespie, J.H. Virus induced congenital anomalies of the bovine fetus. I. Cerebellar degeneration (hypoplasia), ocular lesions and fetal mummification following experimental infection with bovine viral diarrhea-mucosal disease virus. Cornell Vet., 1973, 63:536-560.

2110. Scott, F.W., Kahrs, R.F. and Parsonson, I.M. A cytopathogenic strain of bovine viral diarrhea-mucosal disease virus isolated from a bovine fetus. Cornell Vet., 1972, 62:74-84.

2111. Scott, G.B., Buck, B.E., Leterman, J.G., Bloom, F.L. and Parks, W.P. Acquired immunodeficiency syndrome in infants. New England Journal of Medicine, 1984, 310:76-81.

2112. Scott, G.R. Symposium: the smallest stowaways. 1.--African swine fever. Vet. Rec., 1965, 77:1421-1427.

2113. Scott, H.D. and Byrne, E.B. Exposure of susceptible pregnant women to rubella vaccines. Serologic findings during the Rhode Island immunization campaign. J.A.M.A., 1971, 215:609-612.

2114. Scott, R., Scott, M. and Toms, G.L. Cellular and antibody response to respiratory syncytial (RS) virus in human colostrum, maternal blood, and cord blood. J. Med. Virol., 1981, 8:55-56.

2115. Sekeles, E. and Ornoy, A. Osseous manifestations of gestational rubella in young human fetuses. Am. J. Obstet. Gynecol., 1975, 122:307-312.

2116. Sellars, S.L. The glue ear and rubella risk children. Proc. R. Soc. Med., 1969, 62:111-112.

2117. Selzer, G. Virus isolation, inclusion bodies and chromosomes in a rubella-infected human embryo. Lancet, 1963, 336-337.

2118. Selzer, G. Rubella in pregnancy. S. Afr. J. Obstet. Gynaecol., 1964, 2:5-9.

2119. Selzer, G. Transplacental infection of the mouse fetus by Coxsackie viruses. Isr. J. Med. Sci., 1969, 5:125-127.

2120. Serdula, M.K., Marks, J.S., Herrmann, K.L., Orenstein, W.A., Hall, A.D. and Bomgaars, M.R. Therapeutic abortions following rubella infection in pregnancy: the potential impact on the incidence of congenital rubella syndrome. Am. J. Public Health, 1984, 74:1249-1251.

2121. Serdula, M.K., Marks, J.S., Remington, P.L., Ibara, C.M. and White, M.C. Premarital rubella screening program: from identification to vaccination of susceptible women in the state of Hawaii. Public Health Rep., 1986, 101:329-333.

2122. Seth, P. Rubella virus, pregnancy and fetus. Indian J. Pediatr., 1985, 52:386-389.

2123. Sever, J. and White, L.R. Intrauterine viral infections. Ann. Rev. Med., 1968, 19:471-486.

2124. Sever, J.L. Rubella epidemiology and vaccines. Sight Sav. Rev., 1967, 37:68-72.

2125. Sever, J.L. The epidemiology of rubella. Arch. Ophthalmol., 1967, 77:427-429.

2126. Sever, J.L. The prevention of mental retardation through control of infectious diseases. Public Health Service Publication, 1968, 1692:37.

2127. Sever, J.L. Viruses and the fetus. Int. J. Gynecol. Obstet., 1970, 8:763.

2128. Sever, J.L. Persistent central nervous system infection by a virus: rubella. Res. Publ. Assoc. Res., Nerv. Ment. Dis., 1971, 49:365-372.

2129. Sever, J.L. Persistent central nervous system infection by a virus: rubella. Res. Publ. Assoc. Res., Nerv. Ment. Dis., 1971, 49:365-372.

2130. Sever, J.L. Effects of infections on pregnancy risk. Clin. Obstet. Gynecol., 1973, 16:225-234.

2131. Sever, J.L. Present status of vaccines for rubella. Arch. Otolaryngol., 1973, 98:265-268.

2132. Sever, J.L. Infectious agents as teratogens. In: Shepard, T.H. et al., ed. Methods for detection of environmental agents that produce congenital defects. Amsterdam, North-Holland, 1975.

2133. Sever, J.L. Viral infections in pregnancy. Clin. Obstet. Gynecol., 1978, 21:477-487.

2134. Sever, J.L., Ellenberg, J.H., Ley, A. and Edmonds, D. Incidence of clinical infections in a defined population of pregnant women. Prog. Clin. Biol. Res., 1985, 163:317-326.

2135. Sever, J.L., Fuccillo, D.A., Gilkeson, M.R., Ley, A. and Traub, R. Changing susceptibility to rubella. Obstet. Gynecol., 1968, 32:365-369.

2136. Sever, J.L., Hardy, J.B., Nelson, K.B. and Gilkeson, M.R. Rubella in the collaborative perinatal research study. II. Clinical and laboratory findings in children through 3 years of age. Am. J. Dis. Child, 1969, 118:123-132.

2137. Sever, J.L. and London, W.T. Viruses and embryos. Teratology, 1969, 2:39-46.

2138. Sever, J.L., Meier, G.W., Windle, W.F., Schiff, G.M., Monif, G.R. and Fabiyi, A. Experimental rubella in pregnant rhesus monkeys. J. Infect. Dis., 1966, 116:21-26.

2139. Sever, J.L. and Monif, G. Limited persistence of virus in congenital rubella. Am. J. Dis. Child, 1965, 110:452-454.

2140. Sever, J.L., Nelson, K.B. and Gilkeson, M.R. Rubella epidemic, 1964: effect on 6,000 pregnancies. Am. J. Dis. Child, 1965, 110:395-407.

2141. Sever, J.L. and White, L.R. Intra-uterine viral infections. Anna. Rev. Med., 1968, 19:471.

2142. Sever, L.E. Influenza and congenital malformations of the central nervous system. Lancet, 1972, 1:910-911.

2143. Shabot, J.M., Jaynes, C., Little, H.M. Jr., Alperin, J.B. and Snyder, N. Viral hepatitis in pregnancy with disseminated intravascular coagulationa and hypoglycemia. South Med. J., 1978, 71:479-481.

2144. Shah, K., Daniel, R., Madden, D. and Stagno, S. Serological investigation of BK papovavirus infection in pregnant women and their offspring. Infect. Immun., 1980, 30:29-35.

2145. Shah, K., Kashima, H., Polk, B.F., Shah, F., Abbey, H. and Abramson, A. Rarity of cesarean delivery in cases of juvenile-onset respiratory papillomatosis. Obstet. Gynecol., 1986, 68:795-799.

2146. Shah, K.V., Rangan, S.R., Reissig, M., Daniel, R.W. and Belhan, F.Z. Congenital transmission of a papovavirus of the stump-tailed macaque. Science, 1977, 195:404-406.

2147. Shahin, B., Papadopoulou, Z. and Jenis, E.H. Congenital nephrotic syndrome associated with congenital toxoplasmosis. J. Pediat., 1974, 85:366.

2148. Shalev, E. and Bassan, H.M. Viral hepatitis during pregnancy in Israel. Int. J. Gynaecol. Obstet., 1982, 20:73-78.

2149. Shanmugam, J., Balakrishnan, V., Venugopalan, P. and Sukumaran, C. Prevalence of hepatitis B surface antigen in blood donors and pregnant women from southern Kerala. Indian J. Med. Res., 1978, 68:91-96.

2150. Sharma, R. and Jagdev, D.K. Congenital smallpox. Scand. J. Infect. Dis., 1971, 3:245-247.

2151. Sharpe, H.B. Intranuclear inclusion bodies in the intestinal epithelium of pig foetuses experimentally infected with porcine adenovirus. J. Pathol. Bacteriol., 1967, 93:353-355.

2152. Shaw, E.B. Gonorrheal ophthalmia neonatorum. Pediatrics, 1973, 52:281-282.

2153. Shaw, E.B. Autarcesis against measles in early infancy. J. Infect. Dis., 1979, 140:1015-1016.

2154. Shaw, G.M., Harper, M.E., Hahn, B.H., Epstein, L.G., Gajdusek, D.C., Price, R.W., Navia, B.A., Petito, C.K., O'Hara, C.J., Groopman, J.E., Cho, E.S., Oleske, J.M., Wong-Staal, F. and Gallo, R.C. `HTLV-III infection in brains of children and adults with AIDS encephalopathy. 1985.

2155. Shearer, W.T., Schreiner, R.L., Marshall, R.E. and Barton, L.L. Cytomegalovirus infection in a newborn dizygous twin. J. Pediatr., 1972, 81:1161-1165.

2156. Sheinbergas, M.M. Hydrocephalus due to prenatal infection with the lymphocytic choriomeningitis virus. Infection, 1976, 4:185-191.

2157. Sheinbergas, M.M., Kilchavskiene, V.V. and Tulevichiene, J.P. Prenatal lymphocytic choriomeningitis (LCM): three new cases. Infection, 1984, 12:105-106.

2158. Sheinis, M., Sarov, I., Maor, E. and Gorodischer, R. Severe neonatal rubella following maternal infection. Pediatr. Infect. Dis., 1985, 4:202-203.

2159. Shellam, G.R. and Flexman, J.P. Genetically determined resistance to murine cytomegalovirus and herpes simplex virus in newborn mice. J. Virol., 1986, 58:152-156.

2160. Sheridan, M.D. Final report of a prospective study of children whose mothers had rubella in early pregnancy. Br. Med. J., 1964, 2:536-539.

2161. Shimizu, T. and Kawashima, Y. An outbreak of mare abortion in Japan due to infection with equine rhinopneumonitis virus. Bull. Off. Int. Epizoot., 1968, 70:251-256.

2162. Shimizu, Y., Nakano, K., Inui, S. and Murase, N. Isolation of a strain of infectious bovine rhinotracheitis virus from aborted fetuses. Natl. Inst. Anim. Health Q. (Tokyo), 1972, 12:110-111.

2163. Shin, Y.H. and Glass, L. and Evans, H.E. The TORCH syndrome. Pediatr. Ann., 1976, 5:106-113.

2164. Shinozaki, T., Saito, K. and Shiraki, K. HBsAg-positive giant cell hepatitis with cirrhosis in a 10-month-old infant. Arch. Dis. Child, 1981, 56:64-66.

2165. Shiraki, K., Yoshihara, N., Kawana, T., Yasui, H. and Sakurai, M. Hepatitis B surface antigen and chronic hepatitis in infants born to asymptomatic carrier mothers. Am. J. Dis. Child, 1977, 131:644-647.

2166. Shlian, D.M. Screening and immunization of rubella-susceptible women. Experience in a large, prepaid medical group. J.A.M.A., 1978, 240:662-663.

2167. Shope, R.E. Jr. Bovine virus diarrhea: infectious bovine rhinotracheitis complex. J. Dairy Sci., 1970, 53:619-621.

2168. Shore, S.L., Milgrom, H., Wood, P.A. and Nahmias, A.J. Antibody-dependent cellular cytotoxicity to target cells infected with herpes simplex viruses: functional adequacy in the neonate. Pediatrics, 1977, 59:22-28.

2169. Shulman, N.R. Hepatitis-associated antigen. Am. J. Med., 1970, 49:669-692.

2170. Shwer, M. and Moosa, A. The effects of hepatitis A and B in pregnancy on mother and fetus. S. Afr. Med. J., 1978, 54:1092-1095.

2171. Sibille, G., Sarda, P., Jalaguier, J., Eliaou, J.F., Montoya, F. and Bonnet, H. Reinfection after rubella in newborns with malformations. J. Genet. Hum., 1986, 34:305-312.

2172. Sieber, O.F. Jr., Fulginiti, V.A., Brazie, J. and Umlauf, H.J. Jr. In utero infection of the fetus by herpes simplex virus. J. Pediatr., 1966, 69:30-34.

2173. Siegel, M. Congenital malformations following chickenpox, measles, mumps, and hepatitis. Results of a cohort study. J.A.M.A., 1973, 226:1521-1524.

2174. Siegel, M. Unresolved issues in the first five years of the rubella immunization program. Am. J. Obstet. Gynecol., 1976, 124:327-332.

2175. Siegel, M. and Fuerst, H.T. Low birth weight and maternal virus diseases. A prospective study of rubella, measles, mumps, chickenpox, and hepatitis. J.A.M.A., 1966, 197:88.

2176. Siegel, M. and Fuerst, H.T. Low birth weight and maternal virus diseases: a prospective study of rubella, measles, mumps, chickenpox and hepatitis. J.A.M.A., 1966, 197:680.

2177. Siegel, M., Fuerst, H.T. and Duggan, W. Rubella in mother and congenital cataracts in child. Comparative data in periods with and without epidemics from 1957 to 1964. J.A.M.A., 1968, 203:632-636.

2178. Siegel, M., Fuerst, H.T. and Guinee, V.F. Rubella epidemicity and embryopathy. Results of a long-term study. Am. J. Dis. Child, 1971, 121:469-473.

2179. Siegel, M., Fuerst, H.T. and Peress, N.S. Fetal mortality in maternal rubella. Results of a prospective study from 1957 to 1964. Am. J. Obstet. Gynecol., 1966, 96:247-253.

2180. Siegel, M. and Greenberg, M. Incidence of poliomyelitis in pregnancy, its relation to maternal age, parity and gestational period. N. Engl. J. Med., 1955, 253:841.

2181. Siegel, M. and Greenberg, M. Poliomyelitis in pregnancy: Effect on fetus and newborn infant. J. Pediat., 1956, 49:280.

2182. Siegel, M. and Greenberg, M. Virus diseases in pregnancy and their effect on the fetus. Am. J. Obstet. Gynecol., 1963, 77:620.

2183. Siennicki, W., Oles, A. and Kocielska, W. The persistence of complement fixing and haemagglutination-inhibition antibodies after smallpox infection and after smallpox vaccination. J. Hyg. Epidemiol. Microbiol. Immunol., 1967, 11:27-31.

2184. Silver, H.K. and Dixon, M.S. Jr. Congenital toxoplasmosis; report of case with cataract, atypical vasopressin-sensitive diabetes insipidus, and marked eosinophilia. Amer. J. Dis. Child, 1954, 88:84-91.

2185. Silverman, S.R. Deafness and rubella: infants in the `60s, adults in the `80s. Am. Ann. Deaf., 1980, 125:963-967.

2186. Silverstein, A.M., Parshall, C.J. Jr. and Uhr, J.W. Immunologic maturation in utero: kinetics of the primary antibody response in the fetal lamb. Science, 1966, 154:1675-1677.

2187. Simon, M.W. and McCracken, G.H. Neonatal cytomegalovirus infection. J. Ky. Med. Assoc., 1986, 84:345-350.

2188. Singer, D.B., Rudolph, A.J., Rosenberg, H.S., Rawls, W.E. and Boniuk, M. Pathology of the congenital rubella syndrome. J. Pediatr., 1967, 71:665-675.

2189. Singh, D.S., Balasubramaniam, M., Krishnaswami, S. and Chandrasekar, S. Viral hepatitis during pregnancy. J. Indian Med. Assoc., 1979, 73:90-92.

2190. Singleton, E.B., Rudolph, A.J., Rosenberg, H.S. and Singer, D.B. The roentgenographic manifestations of the rubella syndrome in newborn infants. Am. J. Roentgenol. Radium. Ther. Nucl. Med., 1966, 97:82-91.

2191. Sinha, S.K. An overview of significant research on viral infections of the foetus associated with congenital defects and mental deficiencies. J. Ment. Defic. Res., 1979, 23:207-212.

2192. Skaug, K., Ulstrup, J.C. and Stray-Pedersen, B. A rubella epidemic in an unvaccinated pregnant population. I. Screening methods and serological results. Scand. J. Infect. Dis., 1981, 13:11-15.

2193. Skidmore, S.J., Boxall, E.H. and Ala, F. A case report of post-transfusion hepatitis A. J. Med. Virol., 1982, 10:223.

2194. Skinhj, P. Infection with hepatitis B virus in infancy. A longitudinal study of 8 cases. Arch. Dis. Child, 1978, 53:746-748.

2195. Skinhoj, P. Infectiosity of HB Ag carriers? Gastroenterology, 1974, 66:1102.

2196. Skinhoj, P., Olesen, H., Cohn, J. and Mikkelsen, M. Hepatitis-associated antigen in pregnant women. Acta. Pathol. Microbiol. Scand. [B], 1972, 80:362-366.

2197. Skinhoj, P., Sardemann, H., Cohn, J., Mikkelsen, M. and Olesen, H. Hepatitis associated antigen (HAA) in pregnant women and their newborn infants. Am. J. Dis. Child, 1972, 123:380-381.

2198. Smith, A.W., Akers, T.G., Madin, S.H. and Vedros, N.A. San Miguel sea lion virus isolation, preliminary characterization and relationship to vesicular exanthema of swine virus. Nature, 1973, 244:108-110.

2199. Smith, C.A. and Halse, L. Ophthalmia neonatorum. Public Health Rep., 1955, 70:462-470.

2200. Smith, G.H. Role of the milk agent in disappearance of mammary cancer in C3H/StWi mice. J. Natl. Cancer Inst., 1966, 36:685-701.

2201. Smith, I.M., Girard, A., Corner, A.H. and Mitchell, D. The fluorescent antibody technique in the diagnosis of equine rhinopneumonitis virus abortion. Can. J. Comp. Med., 1972, 36:303-308.

2202. Smith, M.W., Miller, R.B., Svoboda, I. and Lawson, K.F. Efficacy of an intranasal infectious bovine rhinotracheitis vaccine for the prevention of abortion in cattle. Can. Vet. J., 1978, 19:63-71.

2203. Smith, R.N. and Hanna, L. Herpesvirus infections in pregnancy: a comparison of neutralizing antibody titers in mothers and their infants. Am. J. Obstet. Gynecol., 1974, 119:314-319.

2204. Smithells, R.W. New concepts of rubella embryopathy. Dev. Med. Child Neurol., 1966, 8:79-80.

2205. Smithells, R.W., Sheppard, S., Holzel, H. and Dickson, A. National Congenital Rubella Surveillance Programme 1 July 1971-30 June 1984. Br. Med. J. [Clin. Res.], 1985, 291:40-41.

2206. Smithells, R.W., Sheppard, S., Marshall, W.C. and Peckham, C.S. Congenital rubella and diabetes melitus. Lancet, 1978, 1:439.

2207. Smithwick, E.M. and Go, S.C. Hepatitis-associated antigen in cord and maternal sera. Lancet, 1970, 2:1080-1081.

2208. Smithwick, E.M., Pascual, E. and Go, S.C. Hepatitis-associated antigen: a possible relationship to premature delivery. J. Pediatr., 1972, 81:537-540.

2209. Snodgrass, D.R., Fahey, K.J., Wells, P.W., Campbell, I. and Whitelaw, A. Passive immunity in calf rotavirus infections: maternal vaccination increases and prolongs immunoglobulin G1 antibody secretion in milk. Infect. Immun., 1980, 28:344-349.

2210. Snowdon, W.A., Parsonson, I.M. and Broun, M.L. The reaction of pregnant ewes to inoculation with mucosal disease virus of bovine origin. J. Comp. Pathol., 1975, 85:241-251.

2211. Snowe, R.J. and Wilfert, C.M. Epidemic reappearance of gonococcal ophthalmia neonatorum. Pediatrics, 1973, 51:110-114.

2212. Snydman, D.R. Hepatitis in pregnancy. N. Engl. J. Med., 1985, 313:1398-1401.

2213. Sobey, W.R. and Conolly, D. Myxomatosis: passive immunity in the offspring of immune rabbits (Oryctolagus cuniculus) infested with fleas (Spilopsyllus cuniculi Dale) and exposed to myxoma virus. J. Hyg. (Camb.), 74:43-55.

2214. Soike, K. Coxsackie B-3 virus infection in the pregnant mouse. J. Infect. Dis., 1967, 117:203.

2215. Soike, K.F., Iatropoulis, M. and Siegl, G. Infection of newborn and fetal hamsters induced by inoculation of Lulll parvovirus. Arch. Virol., 1976, 51:235-241.

2216. Solov'ev, V.D., Khesin, Y.A. and Gutman, N.R. On the disruption of intra-uterine development in mice in pregnant females inoculated with Coxsackie viruses. Vopr. Virusol., 1970, 15:60.

2217. Soothill, J.F., Hayes, K. and Dudgeon, J.A. The immunoglobulins in congenital rubella. Lancet, 1966, 1:1385-1388.

2218. South, J. Maternal influenza and perinatal mortality. Br. Med. J., 1972, 2:464-465.

2219. South, M.A. and Sever, J.L. Teratogen update: the congenital rubella syndrome. Teratology, 1985, 31:297-307.

2220. South, M.A., Tompkins, W.A., Morris, C.R. and Rawls, W.E. Congenital malformation of the central nervous system associated with genital type (type 2) herpesvirus. J. Pediatr., 1969, 75:13-18.

2221. Spector, S.A. and Straube, R.C. Protean manifestations of perinatal enterovirus infections. West J. Med., 1983, 138:847-851.

2222. Spence, M.R. Inapparent sexual transmission of herpes simplex virus type II infection with subsequent neonatal death: case report. Am. J. Obstet. Gynecol., 1978, 130:590-591.

2223. Spence, M.R., Davidson, D.E., Dill, G.S. Jr., Boonthai, P. and Sagartz, J.W. Rabies exposure during pregnancy. Am. J. Obstet. Gynecol., 1975, 123:655-656.

2224. Sperling, D.R. and Verska, J.J. Rubella syndrome. Cardiovascular manifestation and surgical therapy in infants. Calif. Med., 1966, 105:340-344.

2225. Spertzel, R.O., Crabbs, C.L. and Vaughn, R.E. Transplacental transmission of Venezuelan equine encephalomyelitis virus in mice. Infect. Immun., 1972, 6:339-343.

2226. Spieker, J.O. and Yuill, T.M. Herpesvirus sylvilagus in cottontail rabbits: evidence of shedding but not transplacental transmission. J. Wildl. Dis., 1977, 13:85-89.

2227. Spirer, Z., Levy, J., Bogair, N. and David, M.P. Nitroblue-tetrazolium reduction test in pregnancy. Am. J. Obstet. Gynecol., 1974, 118:1089-1092.

2228. Squartini, F., Bucciarelli, E. and Bolis, G.B. Presence and transmission of mammary tumor virus and leukemia virus in the BALB-cfR111 mouse. J. Natl. Cancer Inst., 1974, 53:137-150.

2229. Srabstein, J.C., Morris, N., Larke, R.P., DeSa, D.J., Castelino, B.B. and Sum, E. Is there a congenital varicella syndrome? J. Pediatr., 1974, 84:239-243.

2230. Srensen, K.J. and Askaa, J. Fetal infection with porcine parvovirus in herds with reproductive failure. Acta. Vet. Scand., 1981, 22:162-170.

2231. Srensen, K.J., Askaa, J. and Dalsgaard, K. Elevated immunoglobulin levels in the diagnosis of intrauterine parvovirus infections in pigs. Acta. Vet. Scand., 1980, 21:152-154.

2232. St. Clair, E.H., Diliberte, J.H. and O'Brien, M.L. Observations of an infant born to a mother with botulism. J. Pediat., 1975, 87:658.

2233. St. Geme, J.W. Jr., Bailey, S.R., Koopman, J.S., Oh, W., Hobel, C.J. and Imagawa, D.T. Neonatal risk following late gestational genital herpesvirus hominis infection. Am. J. Dis. Child, 1975, 129:342-343.

2234. St. Geme, J.W. Jr., Noren, G.R. and Adams, P. Jr. Proposed embryopathic relation between mumps virus and primary endocardial fibroelastosis. N. Engl. J. Med., 1966, 275:339-347.

2235. St. Geme, J.W. Jr., Peralta, H., Farias, E., Davis, C.W. and Noren, G.R. Experimental gestational mumps virus infection and endocardial fibroelastosis. Pediatrics, 1971, 48:821-826.

2236. St. Geme, J.W. Jr., Peralta, H. and VanPelt, L.F. Intrauterine infection of the rhesus monkey with mumps virus: abbreviated viral replication in the immature fetus as an explanation for split immunologic recognition after birth. J. Infect. Dis., 1972, 126:249-256.

2237. St. Geme, J.W. Jr. and VanPelt, L.F. Fetal and postnatal growth retardation associated with gestational mumps virus infection of the rhesus monkey. Lab. Anim. Sci., 1974, 24:895-899.

2238. St. Geme, J.W. Jr., VanPelt, L.F., Yamauchi, T. and Noren, G.R. Transplacental mumps virus infection. Am. J. Obstet. Gynecol., 1976, 125:280.

2239. Stadlan, E.M. and Sung, J.H. Congenital rubella encephalopathy. J. Neuropathol. Exp. Neurol., 1967, 26:115.

2240. Stagno, S., Cloud, G., Pass, R.F., Britt, W.J. and Alford, C.A. Factors associated with primary cytomegalovirus infection during pregnancy. J. Med. Virol., 1984, 13:347-353.

2241. Stagno, S., Pass, R.F. and Alford, C.A. Perinatal infections and maldevelopment. Birth Defects, 1981, 17:31-50.

2242. Stagno, S., Pass, R.F., Cloud, G., Britt, W.J., Henderson, R.E., Walton, P.D., Veren, D.A., Page, F. and Alford, C.A. Primary cytomegalovirus infection in pregnancy. Incidence, transmission to fetus, and clinical outcome. J.A.M.A., 1986, 256:1904-1908.

2243. Stagno, S., Pass, R.F. and Dworsky, M.E. Congenital cytomegalovirus infection, the relative importance of primary and recurrent maternal infection. N. Engl. J. Med., 1982, 306:945-949.

2244. Stagno, S., Pass, R.F. and Dworsky, M.E. Maternal cytomegalovirus infection and perinatal transmission. Clin. Obstet. Gynecol., 1982, 25:563-576.

2245. Stagno, S., Pass, R.F., Dworsky, M.E. and Alford, C.A. Jr. Maternal cytomegalovirus infection and perinatal transmission. Clin. Obstet. Gynecol., 1982, 25:563-576.

2246. Stagno, S., Pass, R.F., Dworsky, M.E., Britt, W.J. and Alford, C.A. Congenital and perinatal cytomegalovirus infections: clinical characteristics and pathogenic factors. Birth Defects, 1984, 20:65-85.

2247. Stagno, S., Pass, R.F., Dworsky, M.E., Henderson, R.E., Moore, E.G., Walton, P.D. and Alford, C.A. Congenital cytomegalovirus infection: The relative importance of primary and recurrent maternal infection. N. Engl. J. Med., 1982, 306:945-949.

2248. Stagno, S., Reynolds, D.W. and Amos, C.S. Auditory and visual defects resulting from symptomatic and subclinical congenital cytomegaloviral and toxoplasma infections. Pediatrics, 1977, 59:669-678.

2249. Stagno, S., Reynolds, D.W., Huang, E.S., Thames, S.D., Smith, R.J. and Alford, C.A. Congenital cytomegalovirus infection. N. Engl. J. Med., 1977, 296:1254-1258.

2250. Stagno, S., Reynolds, D.W., Lakeman, A., Charamella, L.J. and Alford, C.A. Congenital cytomegalovirus infection: consecutive occurrence due to viruses with similar antigenic compositions. Pediatrics, 1973, 52:788-794.

2251. Stagno, S., Reynolds, D.W. and Smith, R.J. Use of isolated nuclei in the indirect fluorescent-antibody test for human cytomegalovirus infection: comparison with microneutralization, anticomplement, and conventional indirect fluorescent-antibody assays. J. Clin. Microbiol., 1978, 7:486-489.

2252. Stagno, S., Tinker, M.K., Elrod, C., Fuccillo, D.A., Cloud, G. and OBeirne, A.J. Immunoglobulin M antibodies detected by enzyme-linked immunosorbent assay and radioimmunoassay in the diagnosis of cytomegalovirus infections in pregnant women and newborn infants. J. Clin. Microbiol., 1985, 21:930-935.

2253. Stagno, S. and Whitley, R.J. Herpesvirus infections of pregnancy. Part I: Cytomegalovirus and Epstein-Barr virus infections. N. Engl. J. Med., 1985, 313:1270-1274.

2254. Stagno, S. and Whitley, R.J. Herpesvirus infections of pregnancy. Part II: Herpes simplex virus and varicella-zoster virus infections. N. Engl. J. Med., 1985, 313:1327-1330.

2255. Stallman, N.D., Allan, B.C. and Sutherland, C.J. Prolonged rubella IgM antibody response. Med. J. Aust., 1974, 2:629-631.

2256. Stanley, F.J., Burgar, P.J., Fong, N.W. and Milroy, H.M. Congenital rubella syndrome in Western Australia. Aust. Paediatr. J., 1985, 21:111-114.

2257. Starr, J.G. Cytomegalovirus infection in pregnancy. N. Engl. J. Med., 1970, 282:50-51.

2258. Starr, J.G., Bart, R.D. Jr. and Gold, E. Inapparent congenital cytomegalovirus infection: Clinical and epidemiologic characteristics in early infancy. New Eng. J. Med., 1970, 282:1075-1078.

2259. Staub, H.P. Thrombocytopenic Purpura and intra-uterine growth retardation of the newborn following maternal rubella. Minn. Med., 1967, 50:1079-1081.

2260. Staub, H.P. Rubella as a possible cause of diastrophic dwarfism. J. Pediatr., 1970, 77:1096-1097.

2261. Stauber, E. and Card, C. Experimental intraamnionic exposure of bovine fetuses with subgroup 2, type 7 Adenovirus. Can. J. Comp. Med., 1978, 42:466-472.

2262. Stein, P.J. and Siciliano, A. Necrotizing herpes simplex viral infection of the cervix during pregnancy. A mimic of squamous cell carcinoma. Am. J. Obstet. Gynecol., 1966, 94:249-252.

2263. Stephenson, J.R. Rubella vaccine for adolescents and young adults. J.A.M.A., 1970, 213:1040.

2264. Sterk, V.V., Prywes, R., Davies, A.M., Ever-Hadani, P. and Lilos, P. Vomiting during pregnancy in Jerusalem women. Isr. J. Med. Sci., 1971, 7:1248-1255.

2265. Stern, H. Cytomegalovirus: a cause of persistent latent infection. J. Clin. Pathol. (R. Coll. Pathol.), 1972, 34-38.

2266. Stern, H. Intrauterine infection with cytomegalovirus. Proc. R. Soc. Med., 1975, 68:367-368.

2267. Stern, H. Cytomegalovirus infection. Nurs. Times, 1977, 73:190-192.

2268. Stern, H. Intrauterine and perinatal cytomegalovirus infections. J. Antimicrob. Chemother., 1979, 5:81-85.

2269. Stern, H., Elek, S.D., Booth, J.C. and Fleck, D.G. Microbial causes of mental retardation. The role of prenatal infections with cytomegalovirus, rubella virus, and toxoplasmosis. Lancet, 1969, 2:443-448.

2270. Stern, H., Elek, S.D., Booth, J.C. and Fleck, D.G. Microbial causes of mental retardation. The role of prenatal infections with cytomegalovirus, rubella virus and toxoplasma. Lancet, 1969, 2:443-448.

2271. Stern, H., Hannington, G., Booth, J. and Moncrieff, D. An early marker of fetal infection after primary cytomegalovirus infection in pregnancy. Br. Med. J. [Clin. Res.], 1986, 292:718-720.

2272. Stern, H. and Tucker, S.M. Prospective study of cytomegalovirus infection in pregnancy. Br. Med. J., 1973, 2:268-270.

2273. Stevens, C.E. Viral hepatitis in pregnancy: the obstetrician's role. Clin. Obstet. Gynecol., 1982, 25:577-584.

2274. Stewart, W.C., Carbrey, E.A. and Kresse, J.I. Transplacental hog cholera infection in immune sows. Am. J. Vet. Res., 1972, 33:791-798.

2275. Stewart, W.C., Carbrey, E.A. and Kresse, J.I. Transplacental hog cholera infection in susceptible sows. Am. J. Vet. Res., 1973, 34:637-640.

2276. Stewart, W.C., Miller, L.D., Kresse, J.I. and Snyder, M.L. Bovine viral diarrhea infection in pregnant swine. Am. J. Vet. Res., 1980, 41:459-462.

2277. Stewien, K.E., Barbosa, V., DeLima, O.S. and Osiro, K. The influence of maternally derived antibody on the efficacy of further attenuated measles vaccine. Infection, 1978, 6:207-210.

2278. Stiehm, E.R., Ammann, A.J. and Cherry, J.D. Elevated cord macroglobulins in the diagnosis of intrauterine infections. N. Engl. J. Med., 1966, 275:971-977.

2279. Stokes, J.S. Jr. A practical approach to the use of rubella vaccine. Ann. Intern. Med., 1971, 74:136-137.

2280. Stoller, A. Virus-chromosome interaction as a possible cause of cases of Down's syndrome (mongolism) and other congenital anomalies. In: D.M.H. Woolham (ed.) Advances in Teratology, 1968, 3:97.

2281. Stoller, A. and Collmann, R.D. Incidence of infective hepatitis followed by Down's syndrome nine months later. Lancet, 1965, 2:1221-1223.

2282. Stoller, A. and Collmann, R.D. Virus aetiology for Down's syndrome (mongolism). Nature, 1965, 208:903-904.

2283. Stoller, A. and Collmann, R.D. Area relationship between incidences of infectious hepatitis and of the births of children with Down's syndrome nine months later. J. Ment. Defic. Res., 1966, 10:84-88.

2284. Storz, J., Young, S., Carroll, E.J., Bates, R.C., Bowen, R.A. and Keney, D.A. Parvovirus infection of the bovine fetus: distribution of infection, antibody response, and age-related susceptibility. Am. J. Vet. Res., 1978, 39:1099-1102.

2285. Stott, J.L., Lauerman, L.H. and Luedke, A.J. Bluetongue virus in pregnant elk and their calves. Am. J. Vet. Res., 1982, 43:423-428.

2286. Stout, D. Problems of rubella control. N. Engl. J. Med., 1970, 283:819.

2287. Strassburg, M.A., Imagawa, D.T., Fannin, S.L., Turner, J.A., Chow, A.W., Murray, R.A. and Cherry, J.D. Rubella outbreak among hospital employees. Obstet. Gynecol., 1981, 57:283-288.

2288. Strauss, L. and Bernstein, J. Neonatal hepatitis in congenital rubella. A histopathological study. Arch. Pathol., 1968, 86:317-327.

2289. Strauss, M. and Davis, G.L. Viral disease of the labyrinth. I. Review of the literature and discussion of the role of cytomegalovirus in congenital deafness. Ann. Otol. Rhinol. Laryngol., 1973, 82:577-583.

2290. Strauss, M., Krieger, R., Lord, S.G., Tenser, R.B., Abt, A.B. and Maisels, M.J. Clinicopathologic study of cytomegalovirus and congenital hearing loss: a progress report. Trans. Pa. Acad. Ophthalmol. Otolaryngol., 1980, 33:175-182.

2291. Strawn, E.Y. and Scrimenti, R.J. Intrauterine herpes simplex infection. Am. J. Obstet. Gynecol., 1973, 115:581-582.

2292. Streissguth, A.P., Vanderveer, B.B. and Shepard, T.H. Mental development of children with congenital rubella syndrome. A preliminary report. Am. J. Obstet. Gynecol., 1970, 108:391-399.

2293. Strulovici, D. and Copelovici, Y. Isolation of cytomegalic virus from the placenta in a case of inapparent human infection. Rev. Roum. Virol., 1974, 25:265-270.

2294. Stubbings, D.P. and Cameron, I.R. Bovine abortion associated with infectious bovine rhinotracheitis virus infection. Vet. Rec., 1981, 108:101-102.

2295. Stuckless, E.R. Projections for deaf students with maternal rubella: college and other alternatives. Am. Ann. Deaf., 1980, 125:985-992.

2296. Studdert, M.J. and Blackney, M.H. Equine herpesviruses: on the differentiation of respiratory from foetal strains of type 1. Aust. Vet. J., 1979, 55:488-492.

2297. Studdert, M.J., Fitzpatrick, D.R., Horner, G.W., Westbury, H.A. and Gleeson, L.J. Molecular epidemiology and pathogenesis of some equine herpesvirus type 1 (equine abortion virus) and type 4 (equine rhinopneumonitis virus) isolates. Aust. Vet. J., 1984, 61:345-348.

2298. Studdert, M.J., Simpson, T. and Roizman, B. Differentiation of respiratory and abortigenic isolates of equine herpesvirus 1 by restriction endonucleases. Science, 1981, 214:562-564.

2299. Studdert, M.J., Wilks, C.R. and Coggins, L. Antigenic comparisons and serologic survey of equine adenoviruses. Am. J. Vet. Res., 1974, 35:693-699.

2300. Suffin, S.C., Muck, K.B. and Porter, D.D. Vesicular stomatitis virus causes abortion and neonatal death in ferrets. J. Clin. Microbiol., 1977, 6:437-438.

2301. Suffin, S.C., Prince, G.A., Muck, K.B. and Porter, D.D. Immunoprophylaxis of respiratory syncytial virus infection in the infant ferret. J. Immunol., 1979, 123:10-14.

2302. Sugamata, M. and Miura, T. Japanese encephalitis virus infection in fetal mice at different stages of pregnancy I. Stillbirth. Acta. Virol. (Praha.), 1982, 26:279-282.

2303. Sullivan-Bolyai, J., Hull, H.F., Wilson, C. and Corey, L. Neonatal herpes simplex virus infection in King County, Washington. Increasing incidence and epidemiologic correlates. J.A.M.A., 1983, 250:3059-3062.

2304. Sumaya, C.V. and Gibbs, R.S. Immunization of pregnant women with influenza A/New Jersey/76 virus vaccine: reactogenicity and immunogenicity in mother and infant. J. Infect. Dis., 1979, 140:141-146.

2305. Suni, J., Narvanen, A., Wahlstrom, T., Lehtovirta, P. and Vaheri, A. Monoclonal antibody to human T-cell leukemia virus p19 defines polypeptide antigen in human choriocarcinoma cells and syncytiotrophoblasts of first-trimester placentas. Int. J. Cancer, 1984, 33:293-298.

2306. Suni, J., Wahlstrom, T. and Vaheri, A. Retrovirus p30-related antigen in human syncytiotrophoblasts and IgG antibodies in cord-blood sera. Int. J. Cancer, 1981, 28:559-566.

2307. Suptel, E.A. and Maximovich, N.A. The possibility of intra-uterine infection with Coxsackie virus in mice inoculated by different routes. Acta Virol., 1964, 8:46.

2308. Surjus, A. Effets du virus Coxsackie B3 sur la souris gestante et sa transmission transplacentaire. Ann. Inst. Pasteur, 1961, 100:825.

2309. Sussman, M.L., Struass, L. and Hodes, H.L. Fatal Coxsackie group B virus infection in the newborn: report of a case with necropsy findings and a brief review of the literature. Am. Med. Ass. J. Dis. Child, 1973, 97:488.

2310. Sutherst, J.R. and Burke, M. Rubella in pregnancy: a difficult diagnosis. Br. Med. J., 1975, 2:6856.

2311. Sutton, R.N. Rubella. Nurs. Mirror, 1975, 141:63-65.

2312. Sutton, R.N. and Pullen, H.J. Iatrogenic rubella infection. Lancet, 1975, 1:1381.

2313. Svedmyr, A. and Thoren, C. Rubella immunity of pregnant women in Stockholm. Acta. Paediatr. Scand., 1967, 56:405-407.

2314. Swan, C. Rubella in pregnancy as an aetiological agent in congenital malformations, stillbirths, miscarriage and abortion. J. Obstet. Gynaecol. Br. Emp., 1949, 56:341-363.

2315. Swan, C., Tostevin, A.L. and Black, G.H.B. Final observations on congenital defects in infants following infectious diseases during pregnancy with special reference to rubella. Med. J. Aust., 1946, 2:889.

2316. Swan, C., Tostevin, A.L., Moore, B., Mayo, H. and Black, G.H.B. Congenital defects in infants following infectious diseases during pregnancy. Med. J. Aust., 1943, 2:201-210.

2317. Sweeney, P.J. Genital herpes. J. Tenn. Med. Assoc., 1983, 76:24-28.

2318. Sweet, C., Collie, M.H., Toms, G.L. and Smith, H. The pregnant guinea pig as a model for studying influenza infection in utero: infection of foetal tissues in organ culture and in vivo. Br. J. Exp. Pathol., 1977, 58:133.

2319. Sweet, C., Toms, G.L. and Smith, H. The pregnant ferret as a model for studying the congenital effects of influenza virus infection in utero: infection of feotal tissues in organ culture and in vivo. Br. J. Exp. Pathol., 1977, 58:113.

2320. Sweet, R.L. Chlamydia, group B streptococcus, and herpes in pregnancy. Birth, 1985, 12:17-24.

2321. Swift, B.L. and Kennedy, P.C. Experimentally induced infection of in utero bovine fetuses with bovine parainfluenza 3 virus. Am. J. Vet. Res., 1972, 33:57-63.

2322. Swift, B.L. and Trueblood, M.S. Failure to induce fetal infection by inoculation of pregnant immune heifers with bovine parainfluenza 3 virus. J. Infect. Dis., 1973, 127:713-717.

2323. Symonds, D.A. and Driscoll, S.G. Massive fetal ascites, urethral atresia, and cytomegalic inclusion disease. Am. J. Dis. Child, 1974, 127:895-897.

2324. Symons, M.J. and Jack, I. Lymphocyte viraemia in congenital rubella. Lancet, 1968, 2:953-954.

2325. Szeifert, G., Csecsei, K., Toth, Z. and Papp, Z. Prenatal diagnosis of ascites caused by cytomegalovirus hepatitis. Acta. Paediatr. Hung., 1985, 26:311-316.

2326. Szmuness, W., Prince, A.M., Hirsch, R.L. and Brotman, B. Familial clustering of hepatitis B infection. N. Engl. J. Med., 1973, 289:1162-1166.

# T

2327. Tabsh, K.M., Brinkman, C.R. III, Nuwayhid, B.S. and Sarti, D.A. Serial ultrasound evaluation of an Rh-isoimmunized patient with hepatitis. Obstet. Gynecol., 1981, 57:121-123.

2328. Tafuro, P. and Gurevich, I. Prevention and management of varicella in high-risk individuals. M.C.N., 1984, 9:314-317.

2329. Taguchi, F., Nagaki, D., Saito, M., Haruyama, C. and Iwasaki, K. Transplacental transmission of BK virus in human. Jpn. J. Microbiol., 1975, 19:395-398.

2330. Taina, E., Hanninen, P. and Gronroos, M. Viral infections in pregnancy. Acta. Obstet. Gynecol. Scand., 1985, 64:167-173.

2331. Takahashi, K., Miyakawa, Y., Gotanda, T., Mishiro, S., Imai, M. and Mayumi, M. Shift from free "small" hepatitis B antigen to IgG-bound "large" form in the circulation of human beings and a chimpanzee acutely infected with hepatitis B virus. Gastroenterology, 1979, 77:1193-1199.

2332. Takehara, K., Mitsui, T. and Nakamura, J. (Stillbirth and abortion in hamsters by experimental infection with Japanese encephalitis virus II. Prevention of stillbirth and abortion by immunization [author's transl.]). Virusu, 1975, 25:261-268.

2333. Takeyama, T. Virological studies on experimental infection of pregnant mice with influenza virus. Tohoku J. Exp. Med., 1966, 89:321-340.

2334. Tan, K.L. The third fontanelle. Acta. Paediatr. Scand., 1971, 60:329-332.

2335. Tan, K.L., Wong, T.T., Chan, M.C., Chun, F.Y. and Lam, S.K. Congenital rubella in Singapore. J. Singapore Paediatr. Soc., 1970, 12:111-125.

2336. Tantivanich, S., Auwanich, W. and Tharavanij, S. Infection rates of Cytomegalovirus among Thai pregnant women. Southeast Asian J. Trop. Med. Public Health, 1982, 13:596-600.

2337. Tantivanich, S., Supavej, S., Chityothin, O., Katepunya, A. and Tharavanij, S. Herpes simplex virus and genital infections in Thai pregnant women. Southeast Asian J. Trop. Med. Public Health, 1983, 14:195-198.

2338. Tantivanich, S. and Tharavanij, S. TORCH agents as possible causes of spontaneous abortion in Thai women. J. Med. Assoc. Thai, 1983, 66:443-447.

2339. Tanyi, J., Bajmocy, E., Fazekas, B. and Kaszanyitzky, E.J. Mass abortion caused by infectious rhinotracheitis (IBR/IPV) virus in a beef cattle herd. Acta. Vet. Hung., 1983, 31:135-143.

2340. Taranger, J., Blomberg, J. and Strannegard, O. Intrauterine varicella: a report of two cases associated with hyper-A-immunoglobulinemia. Scand. J. Infect. Dis., 1981, 13:297-300.

2341. Tedder, R.S. Hepatitis B in hospitals. Br. J. Hosp. Med., 1980, 23:266, 275-6, 278-9.

2342. Tejani, N., Klein, S.W. and Kaplan, M. Subclinical herpes simplex genitalis infections in the perinatal period. Am. J. Obstet. Gynecol., 1979, 135:547.

2343. Teodorescu, M., Topciu, V., Plavosin, L., Tudose, N. and Moldovan, E. The influence of smallpox revaccination upon the foetus. Virologie, 1975, 26:137-139.

2344. Terlecki, S., Richardson, C., Done, J.T., Harkness, J.W., Sands, J.J., Shaw, I.G., Winkler, C.E., Duffell, S.J., Patterson, D.S. and Sweasey, D. Pathogenicity for the sheep foetus of bovine virus diarrhoea-mucosal disease virus of bovine origin. Br. Vet. J., 1980, 136:602-611.

2345. Thacker, B.J., Leman, A.D., Hurtgen, J.P., Sauber, T.E. and Joo, H.S. Survey of porcine parvovirus infection in swine fetuses and their dams at a Minnesota abattoir. Am. J. Vet. Res., 1981, 42:865-867.

2346. Thalhammer, O. Congenital toxoplasmosis. Lancet, 1962, 1:23-24.

2347. Theilen, G.H. Continuing studies with transmissible feline fibrosarcoma virus in fetal and newborn sheep. J. Am. Vet. Med. Assoc., 1971, 158:1040.

2348. Theodoridis, A., Giescke, W.H. and DuToit, I.J. Effects of ephemeral fever on milk production and reproduction of dairy cattle. Onderstepoort. J. Vet. Res., 1973, 40:83-91.

2349. Theppisai, U., Chiewsilp, P., Bunyaratvej, S., Siripoonya, P. and Varavithya, W. Hepatitis B surface antigen in asymptomatic carrier mothers and vertical transmission of hepatitis B virus. J. Med. Assoc. Thai, 1984, 67:90-93.

2350 Thiry, E., Saliki, J., Pastoret, P.P., Lambert, A.F. and Ligot, J. Failure to demonstrate infectious bovine rhinotracheitis virus reactivation in parturient cows. Vet. Rec., 1984, 115:248-249.

2351. Thiry, E., Saliki, J., Schwers, A. and Pastoret, P.P. Parturition as a stimulus of IBR virus reativation. Vet. Rec., 1985, 116:599-600.

2352. Thiry, L., Sprecher-Goldberger, S., Bossens, M. and Neuray, F. Immune response to primate oncornaviruses in pre-eclampsia. Lancet, 1978, 1:1268.

2353. Thiry, L., Sprecher-Goldberger, S., Hard, R.C., Bossens, M. and Neuray, F. Expression of retrovirus-related antigen in pregnancy. I. Antigens cross-reacting with simian retroviruses in human foetal tissues and cord blood lymphocytes. J. Reprod. Immunol., 1981, 2:309-322.

2354. Thomas, P.A., Jaffe, H.W., Spira, T.J., Reiss, R., Guerrero, I.C. and Auerbach, D. Unexplained immunodeficiency in children: a surveillance report. Journal of the American Medical Association, 1984, 252:639-644.

2355. Thomas, W.R. Rubella antibody tests in pregnancy. Br. Med. J., 1976, 2:300.

2356. Thompson, J.A. and Glasgow, L.A. Intrauterine viral infection and the cell-mediated immune response. Neurology, 1980, 30:212-215.

2357. Thompson, J.F. and Harvey, V.K. Jr. The "German measles" epidemic in Indiana. Am. J. Obstet. Gynecol., 1966, 95:55-60.

2358. Thompson, K.M. and Tobin, J.O. Isolation of rubella virus from abortion material. Br. Med. J., 1970, 1:264-266.

2359. Thompson, W.M. Jr., Thomas, E.D., Chitwood, L.A., Seely, J.R. and Riley, H.D. Jr. Studies of the rubella syndrome. J. Okla. State Med. Assoc., 1970, 63:97-101.

2360. Thongkrajai, P., Pothinam, S. and Hahnvajanawong, C. A rapid detection of herpes simplex virus by ELISA in asymptomatic pregnant women and neonates. Southeast Asian J. Trop. Med. Public Health, 1986, 17:91-95.

2361. Thorburn, M.J. and Miller, C.G. The pathology of congenital rubella in Jamaica. Arch. Dis. Child, 1967, 42:389-396.

2362. Tidstrom, B. Varicella pneumonia. Treatment with tracheostomy and mechanical endotracheal positive-pressure ventilation. Scand. J. Resp. Dis., 1967, 48:40-44.

2363. Timbury, M.C. Congenital cytomegalovirus infection: a dilemma. Br. Med. J. [Clin. Res.], 1984, 289:712-713.

2364. Ting, C.C., Lavrin, D.H., Shiu, G. and Herberman, R.B. Expression of fetal antigens in tumor cells. Proc. Natl. Acad. Sci. USA, 1972, 69:1664-1668.

2365. Ting, P.L., Chew, W.L. and Tan, B.Y. Viral hepatitis B induced haemolytic anaemia in a patient with normal glucose 6 phosphate dehydrogenase -- a case report. Singapore Med. J., 1984, 25:360-361.

2366. Tobin, J.O. Herpesvirus hominis infection in pregnancy. Proc. R. Soc. Med., 1975, 68:371-374.

2367. Tobin, J.O., Marshall, W.C. and Peckham, C.S. Virus infections. Clin. Obstet. Gynaecol., 1977, 4:479-501.

2368. Tomer, A. and Harel, A. Congenital absence of scalp skin and herpes simplex virus. A case report. Isr. J. Med. Sci., 1983, 19:950-951.

2369. Tondury, G. Embryopathies caused by viruses in pregnancy. Etudes neo-natale, 1953, 2:105.

2370. Tondury, G. Embryopathien; uber die Wirkungsweise (Infektionsweg und Pathogenese (von Viren auf den menschlichen Keimling. (Besrlin: Springer-Verlag), 1962.

2371. Tondury, G., Foukas, M. and Scouteris, A. Prenatal vaccinia. J. Obstet. Gynaecol. Br. Commonw., 1969, 76:47-54.

2372. Tondury, G. and Smith, D.W. Fetal rubella pathology. J. Pediatr., 1966, 68:867-879.

2373. Tondury, G. and Tondury, T.A. Course of infection and pathogenesis of the human embryo. Rev. Suisse Zool., 1972, 79:179.

2374. Tong, M.J., McPeak, C.M., Thursby, M.W., Schweitzer, I.L., Henneman, C.E. and Ledger, W.J. Failure of immune serum globulin to prevent hepatitis B virus infection in infants born to HBsAg-positive mothers. Gastroenterology, 1979, 76:535-539.

2375. Tong, M.J., Sinatra, F.R., Thomas, D.W., Nair, P.V., Merritt R.J. and Wang, D.W. Need for immunoprophylaxis in infants born to HBsAg-positive carrier mothers who are HBeAg negative. J. Pediatr., 1984, 105:945-947.

2376. Tong, M.J., Thursby, M.W., Lin, J.H., Weissman, J.Y. and McPeak, C.M. Studies on the maternal-infant transmission of the hepatitis B virus and HBV infection within families. Prog. Med. Virol., 1981, 27:137-147.

2377. Tong, M.J., Thursby, M., Rakela, J., McPeak, C., Edwards, V.M. and Mosley, J.W. Studies on the maternal-infant transmission of the viruses which cause acute hepatitis. Gastroenterology, 1981, 80:999-1004.

2378. Too, H.L. and Love, R.J. Some epidemiological features and effects on reproductive performance of endemic porcine parvovirus infection. Aust. Vet. J., 1986, 63:50-53.

2379. Toolan, H.W. The picodna viruses. H, RV, and AAV. Int. Rev. Exp. Pathol., 1968, 6:135-180.

2380. Toolan, H.W. Susceptibility of the Syrian hamster to virus infection. Fed. Proc., 1978, 37:2065-2068.

2381. Toole, J.W., Hofstader, S.L. and Ramsay, C.A. Darier's disease and Kaposi's varicelliform eruption. J. Am. Acad. Dermatol., 1979, 1:321-324.

2382. Torphy, D.E., Ray, C.G. and McAlister, R. et al. Herpes simplex virus infection in infants. A spectrum of disease. J. Pediat., 1970, 76:405.

2383. Torrey, E.F. and Peterson, M.R. The viral hypothesis of schizophrenia. Schizophrenia Bull., 1976, 2:136-146.

2384. Townsend, J.J., Baringer, J.R., Wolinsky, J.S., Malamud, J., Mednick, J.P., Panitch, H.S., Scott, R.A.T., Oshiro, L.S. and Cremer, N.E. Progressive rubella panencephalitis, late onset after congenital rubella. N. Engl. J. Med., 1975, 292:990.

2385. Trapp, A.L., Roberts, A.W. and Carter, G.R. Comparison of methods for diagnosing equine rhinopneumonitis virus abortion. Vet. Med. Small Anim. Clin., 1972, 67:895.

2386. Trautwein, G., Hewicker, M., Liess, B., Orban, S. and Grunert, E. Studies on transplacental transmissibility of a bovine virus diarrhoea (BVD) vaccine virus in cattle. III. Occurrence of central nervous system malformations in calves born from vaccinated cows. Zentralbl. Veterinarmed [B], 1986, 33:260-268.

2387. Travers, H. Rubella reinfection. Ann. Intern. Med., 1973, 78:620-621.

2388. Tripathi, B.M. and Misra, N.P. Viral hepatitis with pregnancy. J. Assoc. Physicians India, 1981, 29:463-469.

2389. Trlifajova, J., Benda, R. and Benes, C. Effect of maternal varicella-zoster virus infection on the outcome of pregnancy and the analysis of transplacental virus transmission. Acta. Virol. (Praha.), 1986, 30:249-255.

2390. Trlifajova, J., Hajkova, S., Ryba, M., Sourek, J., Svandova, E. and Seeman, J. The study of the occurrence of precipitation antibodies to varicella-zoster (VZ) virus in healthy and sick children up to one year of age in relation to their transplacental transfer. J. Hyg. Epidemiol. Microbiol. Immunol., 1972, 16:486-492.

2391. Trybus, R.J., Karchmer, M.A., Kerstetter, P.P. and Hicks, W. The demographics of deafness resulting from maternal rubella. Am. Ann. Deaf., 1980, 125:977-984.

2392. Tuboly, S. and Belak, S. Adenovirus-induced immune response in ovine fetuses. Acta. Microbiol. Acad. Sci. Hung., 1980, 27:169-176.

2393. Tucker, M.J. and Stewart, R.B. Intrauterine transmission of Sendai virus in inbred mouse strains. Infect. Immun., 1976, 14:1191-1195.

2394. Turner, A.J., Studdert, M.J. and Peterson, J.E. Equine herpes viruses. 2. Persistence of equine herpesviruses in experimentally infected horses and the experimental induction of abortion. Aust. Vet. J., 1970, 46:90-98.

2395. Turner, G. Recognition of intrauterine growth retardation by considering comparative birth-weights. Lancet, 1971, 2:1123-1124.

2396. Turner, G.C., Field, A.M., Lasheen, R.M., Todd, R.M. and White, G.B. SH (Australia) antigen in early life. J. Clin. Pathol., 1970, 23:826.

2397. Tyms, A.S. Diseases of the fetus and neonate due to human cytomegalovirus: a laboratory perspective. Med. Lab. Sci., 1982, 39:275-286.

2398. Tzipori, S. and Spradbrow, P.B. The effect of bovine ephemeral fever virus on the bovine foetus. Aust. Vet. J., 1975, 51:64-66.

# U

2399. Ueda, K., Hisanaga, S., Nishida, Y. and Shepard, T.H. Low-birth-weight and congenital rubella syndrome: effect of gestational age at time of maternal rubella infection. Clin. Pediatr. (Phila.), 1981, 20:730-733.

2400. Ueda, K., Nishida, Y., Oshima, K., Yoshikawa, H. and Oashi, K. Seven-year follow-up study of rubella syndrome in Ryukyu with special reference to persistence of rubella hemagglutination inhibition antibodies. Jpn. J. Microbiol., 1975, 1975, 19:1810-1815.

2401. Ueda, K., Nishida, Y., Oshima, K., Yoshikawa, H. and Nonaka, S. An explanation for the high incidence of congenital rubella syndrome in Ryukyu. Am. J. Epidemiol., 1978, 107:344-351.

2402. Ueda, S., Okuno, Y., Sangkawibha, N., Jayavasu, J. and Tuchinda, P. Studies on measles in Thailand. I. Seroepidemiological examination. Biken J., 1967, 10:129-133.

2403. Ulstrup, J.C., Skaug, K., Stray-Pedersen, B., Flugsrud, L.B., Halvorsen, S. and Kandhari, M. A rubella epidemic in an unvaccinated pregnant population. 2. Seroepidemiology. Scand. J. Infect. Dis., 1981, 13:83-88.

2404. Umar, M.H. and VanGriensven, L.J. Effects of Rauscher leukemia virus infection on BALB/c mouse embryos. J. Natl. Cancer Inst., 1976, 56:375-380.

2405. Umphenour, N.W., Kemen, M.J. and Coggins, L. Equine infectious anemia: a retrospective study of an epizootic. J. Am. Vet. Med. Assoc., 1974, 164:66-69.

2406. Upfold, L.J. Deafness following rubella in pregnancy. Med. J. Aust., 1970, 1:420-424.

2407. Upfold, L.J. Vaccination, one; maternal rubella, nil. Med. J. Aust., 1985, 143:219.

2408. Upfold, L.J. Vaccination, 2; maternal rubella, O. Med. J. Aust., 1987, 146:119.

2409. Upfold, L.J. Vaccination, one: maternal rubella, nil. Med. J. Aust., 1985, 143:219.

2410. Upfold, L.J. and Isepy, J. Childhood deafness in Australia. Incidence and maternal rubella, 1949-1980. Med. J. Aust., 1982, 2:323-326.

2411. Utley, K., Bromberger, P., Wagner, L. and Schneider, H. Management of primary herpes in pregnancy complicated by ruptured membranes and extreme prematurity: case report. Obstet. Gynecol., 1987, 69:471-473.

# V

2412. Vaccaro, R., Bertotto, A., Zucchetti, P. and Forenza, N. Rashes in congenital rubella. Lancet, 1975, 2:1215.

2413. Vaheri, A. Undesirable properties of rubella vaccines with special reference to their use in women of fertile age. Scand. J. Infect. Dis., 1972, 6:24-27.

2414. Vaheri, A., Vesikari, T., Oker-Blom, N., Seppala, M., Parkman, P.D., Veronelli, J. and Robbins, F.C. Isolation of attenuated rubella-vaccine virus from human products of conception and uterine cervix. N. Engl. J. Med., 1972, 286:1071-1074.

2415. Vaheri, A., Vesikari, T., Oker-Blom, N., Seppala, M., Veronelli, J., Robbins, F.C. and Parkman, P.D. Transmission of attenuated rubella vaccines to the human fetus. A preliminary report. Am. J. Dis. Child, 1969, 118:243-246.

2416. Valenti, W.M. Infection control and the pregnant health care worker. Am. J. Infect. Control, 1986, 14:20-27.

2417. VandePutte, M., Meyer, G. and Sobis, H. Vascular tumors induced by polyoma virus in pregnant rats. J. Natl. Cancer Inst., 1976, 56:517-521.

2418. VanDerMolen, E.J., Vecht, U. and Houwers, D.J. A chronic indurative mastitis in sheep, associated with maedi/visna virus infection. Vet. Q, 1985, 7:112-119.

2419. VanLeengoed, L.A., Vos, J., Gruys, E., Rondhuis, P. and Brand, A. Porcine Parvovirus infection: review and diagnosis in a sow herd with reproductive failure. Vet. Q, 1983, 5:131-141.

2420. VanMeir, J.M., Drogendijk, A.C., Wladimiroff, J.W. and Wielenga, G. Colposcopy in a case of suspected genital herpes infection in early pregnancy. Acta. Obstet. Gynecol. Scand., 1979, 58:221-223.

2421. Vannier, P., Plateau, E. and Tillon, J.P. Congenital tremor in pigs farrowed from sows given hog cholera virus during pregnancy. Am. J. Vet. Res., 1981, 42:135-137.

2422. VanOirschot, J.T. Congenital infections with nonarbo togaviruses. Vet. Microbiol., 1983, 8:321-361.

2423. VanOirschot, J.T. and Gielkens, A.L. In vivo and in vitro reactivation of latent pseudorabies virus in pigs born to vaccinated sows. Am. J. Vet. Res., 1984, 45:567-571.

2424. Varga, A. and Browell, B. Viral inclusion bodies in vaginal smears. Obstet. Gynecol., 1960, 16:441.

2425. Vargas, V., Pedreira, J.D., Esteban, R., Hernandez, J.M., Guardia, J. and Bacardi, R. Materno-fetal transmission of hepatitis A antibody. Acta. Paediatr. Scand., 1980, 69:533.

2426. Varma, R.R. Hepatitis B surface antigen carrier state in neonates. Prophylaxis with large doses of conventional immune human serum globulin. J.A.M.A., 1976, 236:2302-2304.

2427. Varner, M.W., McGuinness, G.A. and Galask, R.P. Rabies vaccination in pregnancy. Am. J. Obstet. Gynecol., 1982, 143:717-718.

2428. Vaucher, Y.E., Ray, C.G., Minnich, L.L., Payne, C.M., Beck, D. and Lowe, P. Pleomorphic, enveloped, virus-like particles associated with gastrointestinal illness in neonates. J. Infect. Dis., 1982, 145:27-36.

2429. Vejtorp, M. and Mansa, B. Rubella IgM antibodies in sera from infants born after maternal rubella later than the 12th week of pregnancy. Scand. J. Infect. Dis., 1980, 12:1-5.

2430. Velasco, A.C., Mateos, M.L., Mas, G., Pedraza, A., Diez, M. and Gutierrez, A. Three-year prospective study of intestinal pathogens in Madrid, Spain. J. Clin. Microbiol., 1984, 20:290-292.

2431. Ventura, A.K., Ehrenkranz, N.J. and Rosenthal, D. Placental passage of antibodies to Dengue virus in persons living in a region of hyperendemic Dengue virus infection. J. Infect. Dis., 1975, 131:S62-68.

2432. Vermeif-Keers, C. Primary congenital aphakia and the rubella syndrome. Teratology, 1975, 11:257-265.

2433. Vernon, M., Grieve, B.J. and Shaver, K. Handicapping conditions associated with the congenital rubella syndrome. Am. Ann. Deaf., 1980, 125:993-997.

2434. Vernon, M. and Hicks, D. Relationship of rubella, herpes simplex, cytomegalovirus, and certain other viral disabilities. Am. Ann. Deaf., 1980, 125:529-534.

2435. Veronelli, J.A., Baublis, J.V., Heyn, R.M. and Hoggatt, J.E. Congenital thrombocytopenic purpura due to intrauterine rubella infection. Mich. Med., 1965, 64:923-928.

2436. Vesikari, T. Rubella antibodies in infants whose mothers had rubella during the second and third trimesters of pregnancy. Scand. J. Infect. Dis., 1971, 3:1-6.

2437. Vesikari, T. Current status of diagnostic procedures for intrauterine rubella and cytomegalovirus infections. Zentralbl. Gynakol., 1983, 105:1051-1056.

2438. Vesikari, T. and Vaheri, A. Rubella: a method for rapid diagnosis of a recent infection by demonstration of the IgM antibodies. Br. Med. J., 1968, 1:221-223.

2439. Vesterinen, E., Savolainen, E.R., Purola, E., Saksela, E. and Leinikki, P. Occurrence of genital herpes simplex and cytomegalovirus infections in pregnancy. Acta. Obstet. Gynecol. Scand., 1977, 56:101-104.

2440. Vianna, N.J. and Polan, A.K. Childhood lymphatic leukemia: prenatal seasonality and possible association with congenital varicella. Am. J. Epidemiol., 1976, 103:321-332.

2441. Villarejos, V.M., Arguedas, J.A., Hernandez, G.C., Buynak, E.B. and Hilleman, M.R. Clinical laboratory evaluation of rubella virus vaccine given to postpartum women without pregnancy preventive. Obstet. Gynecol., 1973, 42:689-695.

2442. Vince, D.J. Prevention of rubella embryopathy. Can. Med. Assoc. J., 1969, 100:777-778

2443. Vince, D.J. The hospital incidence and clinical significance of congenital heart malformations resulting from rubella embryopathy. Can. Med. Assoc. J., 1970, 102:374-376.
100:777-778.

2444. Vince, D.J. The role of rubella in the etiology of supravalvular aortic stenosis. Can. Med. Assoc. J., 1970, 103:1157-1160.

2445. Vogler, C., Kohl, S. and Rosenberg, H.S. Cytomegalovirus infection and fetal death in a twin. A case report. J. Reprod. Med., 1986, 31:207-210.

2446. Vogler, C., Kohl, S. and Rosenberg, H.S. Cytomegalovirus infection and fetal death in a twin. A case report. J. Reprod. Med., 1986, 31:207-210.

2447. Volpi, A., Lakeman, A.D., Pereira, L. and Stagno, S. Monoclonal antibodies for rapid diagnosis and typing of genital herpes infections during pregnancy. Am. J. Obstet. Gynecol., 1983, 146:813-815.

2448. VonBenten, K., Trautwein, G., Richter-Reichhelm, H.B., Liess, B. and Frey, H.R. Experimental Transplacental transmission of hog cholera virus in pigs. III. Histopathological findings in the fetus. Zentralbl. Veterinarmed. [B], 1980, 27:714-724.

2449. VonHerzen, J.L. and Benirschke, K. Unexpected disseminated herpes simplex infection in a newborn. Obstet. Gynecol., 1977, 50:728-730.

2450. Vontver, L.A., Hickok, D.E., Brown, Z., Reid, L. and Corey, L. Recurrent genital herpes simplex virus infection in pregnancy: infant outcome and frequency of asymptomatic recurrences. Am. J. Obstet. Gynecol., 1982, 143:75-84.

2451. Vorst, E.J. and Gaillard, J.L. Vaccinial osteomyelitis in a case of generalized intrauterine virus infection. Pediatr. Pathol., 1983, 1:221-228.

2452. Voyce, M.A. and Hunt, A.C. Congenital tuberculosis. Arch. Dis. Child, 1966, 41:299.

# W


2453. Walker, D.L. and Padgett, B.L. The epidemiology of human polyomaviruses. Prog. Clin. Biol. Res., 1983, 105:99-106.

2454. Walker, I.R. Abortion in early pregnancy. Med. J. Aust., 1985, 142:489.

2455. Wallace, G.D. and Elm, J.L. Jr. Transplacental transmission and neonatal infection with swine influenza virus (Hsw1N1) in swine. Am. J. Vet. Res., 1979, 40:1169-1172.

2456. Waller, B.F., Smith, F.A., Kerwin, D.M. and Roberts, W.C. Fetal rubella 27 years later [clinical conference]. Chest, 1982, 81:735-738.

2457. Wallis, S. and Harvey, D. Intrauterine infection of the fetus. Nurs. Times, 1978, 74:912-916.

2458. Walls, B. Lassa fever and pregnancy. Midwives Chron., 1985, 1168:136-138.

2459. Walpole, I.R. and Grauaug, A. Intra-uterine infection with herpes simplex virus and observed radiological changes. Aust. Paediatr. J., 1979, 15:123-125.

2460. Walters, B.N., Redman, C.W. Bell's palsy and cytomegalovirus mononucleosis in pregnancy. J. R. Soc. Med., 1984, 77:429-430.

2461. Wands, J.R. Viral hepatitis and its effect on pregnancy. Clin. Obstet. Gynecol., 1979, 22:301-311.

2462. Wang, J.T., Dunne, H.W., Griel, L.C., Hokanson, J.F. and Murphy, D.M. Mortality, antibody development, and viral persistence in porcine fetuses incubated in utero with SMEDI (entero-) virus. Am. J. Vet. Res., 1973, 34:785-791.

2463. Wang, Y.J., Kang, Y., Gong, Z.L., Mei, Y.X., Xu, J.Y., Meng, L.X., Zhao, K., Li, X., Hu, Z.H. and Song, Z.Z. Hepatitis B vaccine in neonates: induction of anti-HBs immune response and interruption of maternal HBsAg transmission. Chin. Med. J. [Engl.], 1985, 98:265-270.

2464. Ward, G.M. Bovine viral diarrhea-mucosal disease implicated in a calf with cerebellar hypoplasia and ocular disease. A case report. Cornell Vet., 1971, 61:224-228.

2465. Ward, G.M., Roberts, S.J., McEntee, K. and Gillespie, J.H. A study of experimentally induced bovine viral diarrhea-mucosal disease in pregnant cows and their progeny. Cornell Vet., 1969, 59:525-538.

2466. Ward, P.H., Honrubia, V. and Moore, B.S. Inner ear pathology in deafness due to maternal rubella. Arch. Otolaryngol., 1968, 87:22-28.

2467. Warford, A.L., Levy, R.A., Rekrut, K.A. and Steinberg, E. Herpes simplex virus testing of an obstetric population with an antigen enzyme-linked immunosorbent assay. Am. J. Obstet. Gynecol., 1986, 154:21-28.

2468. Warrell, M.J., Tobin, J.O. and Wald, N.J. Examination for influenza IgA and IgM antibodies in pregnancies associated with fetal neural-tube defects. J. Med. Microbiol., 1981, 14:159-162.

2469. Washburn, T.C., Medearis, D.N. Jr. and Childs, B. Sex differences in susceptibility to infections. Pediat., 1965, 35:57.

2470. Waterson, A.P. Virus infections (other than rubella) during pregnancy. Br. Med. J., 1979, 2:564-566.

2471. Watts, T. Rubella antibodies in a sample of Lusaka mothers. Med. J. Zambia, 1983, 17:109-110.

2472. Way, R.C. Cardiovascular defects and the rubella syndrome. Can. Med. Assoc. J., 1967, 97:1329-1334.

2473. Weber, J.M. Errors with rubella diagnostic serology. Can. Med. Assoc., 1982, 126:901.

2474. Webster, M.H. and Smith, C.S. Congenital abnormalities and maternal herpes zoster. Br. Med. J., 1977, 2:1193.

2475. Weil, M.L., Itabashi, H.H., Cremer, N.E., Oshiro, L.S., Lennette, E.H. and Carnay, L. Chronic progressive panencephalitis due to rubella virus simulating subacute sclerosing panencephalitis. N. Engl. J. Med., 1975, 292:994.

2476. Weinberger, M.M., Maslin, M.W., Asbed, R.A. and Sever, J. Congenital rubella presenting a retarded language development. Am. J. Dis. Child, 1970, 120:125.

2477. Weinstein, L. and Chang, T.W. Prevention of rubella. Pediatrics, 1975, 55:5-6.

2478. Weiss, D.I., Cooper, L.Z. and Green, R.H. Infantile glaucoma. A manifestation of congenital rubella. J.A.M.A., 1966, 195:725-727.

2479. Weller, T.H. The cytomegaloviruses: ubiquitous agents with protean clinical manifestations. New Eng. J. Med., 1971, 285:203-214, 267-274.

2480. Weller, T.H., Alford, C.A. Jr. and Neva, F.A. Changing epidemiologic concepts of rubella, with particular reference to unique characteristics of the congenital infection. Yale J. Biol. Med., 1965, 37:455-472.

2481. Weller, T.H. and Hanshaw, J.B. Virologic and clinical observations on cytomegalic inclusion disease. N. Engl. J. Med., 1962, 266:1233-1244.

2482. Welter, C.J. Comments on vaccines for transmissible gastroenteritis. J. Am. Vet. Med. Assoc., 1972, 160:558-560.

2483. Wentworth, P. Studies on placentae and infants from women vaccinated for smallpox during pregnancy. J. Clin. Pathol., 1966, 19:328-330.

2484. Weppner, W.A. and Coggin, J.H. Jr. Antigenic similarity between plasma membrane proteins of fetal hamster cells and simian virus 40 tumor surface antigens. Cancer Res., 1980, 40:1380-1387.

2485. Wertheim, R.A., Brooks, B.J. Jr., Rodriguez, F.H. Jr., Lesesne, H.R. and Jennette, J.C. Fatal herpetic hepatitis in pregnancy. Obstet. Gynecol., 1983, 62:38s-42s.

2486. Westbury, H.A., Kovesdy, L. and Barton, M.D. Natural intra-uterine infection of foals with equine herpes virus type 1 in Victoria. Aust. Vet. J., 1978, 54:147.

2487. Weston, M. The implications of rubella in pregnancy. Midwife Health Visit Community Nurse, 1984, 20:318-321.

2488. Wetzel, A.M. and Kirz, D.S. Routine hepatitis screening in adolescent pregnancies: is it cost effective? Am. J. Obstet. Gynecol., 1987, 156:166-169.

2489. Whalen, J.P., Winchester, P., Krook, L., ODonohue, N., Dische, R. and Nunez, E. Neonatal transplacental rubella syndrome. Its effect on normal maturation of the diaphysis. Am. J. Roentgenol. Radium. Ther. Nucl. Med., 1974, 121:166-172.

2490. Whipple, D.V. and Dunham, E.C. Congenital syphilis. I. Incidence, transmission and diagnosis. J. Pediat., 1938, 12:386.

2491. White, L.R., Sever, J.L. and Alepa, F.P. Maternal and congenital rubella before 1964: frequency, clinical features, and search for isoimmune phenomena. J. Pediatr., 1969, 74:198-207.

2492. Whitley, R.J., Brasfield, D., Reynolds, D.W., Stagno, S., Tiller, R.E. and Alford, C.A. Protracted pneumonitis in young infants associated with perinatally acquired cytomegaloviral infection. J. Pediatr., 1976, 89:16-22.

2493. Whitley, R.J., Nahmias, A.J., Visintine, A.M., Fleming, C.L. and Alford, C.A. The natural history of herpes simplex virus infection of mother and newborn. Pediatrics, 1980, 66:489-494.

2494. Whitmore, H.L., Zemjanis, R. and Olson, J. Effect of bovine viral diarrhea virus on conception in cattle. J. Am. Vet. Med. Assoc., 1981, 178:1065-1067.

2495. Whitty, R.J. Foetal infections, with special reference to rubella. J. Ir. Med. Assoc., 1967, 60:86-89.

2496. Wilken, H. Clinical picture of prenatal virus infections. Zentralbl. Gynakol., 1976, 98:1559-1564.

2497. Wilken, H. Important clinical aspects of prenatal viral infections. J. Perinat. Med., 1978, 6:257-267.

2498. Wilkinson, R.H. and Heller, R.M. Congenital syphilis: resurgence of an old problem. Pediatrics, 1971, 47:27.

2499. Williams, H.J. and Carey, L.S. Rubella embryopathy. Roentgenologic features. Am. J. Roentgenol. Radium. Ther. Nucl. Med., 1966, 97:92-99.

2500. Williams, K. and Mackenzie, J.S. Influenza infections during pregnancy in the mouse. J. Hyg. (Camb.), 1977, 79:249-257.

2501. Williams, M.T. and Marsland, D.W. Varicella pneumonia at parturition: a non-fatal case report. J. Fam. Pract., 1976, 3:153-154.

2502. Williams, N.M. and Preblud, S.R. Rubella and congenital rubella surveillance, 1983. M.M.W.R. - C.D.C. - Surveill. Summ., 1984, 33:1SS-10SS.

2503. Williamson, A.P. The varicella-zoster virus in the etiology of severe congenital defects. A survey of eleven reported instances. Clin. Pediatr. (Phila.), 1975, 14:558-559.

2504. Williamson, A.P., Blattner, R.J. and Simonsen, L. Cataracts following mumps virus in early chick embryos. Proc. Soc. Exp. Biol. Med., 1957, 96:224.

2505. Williamson, A.P. and South, M.A. Effect of systemic maternal virus infections on the fetus and newborn. Clin. Obstet. Gynecol., 1972, 15:1023-1030.

2506. Williamson, W.D., Desmond, M.M., Lafevers, N., Taber, L., Catlin, F.I. and Weaver, T.G. Symptomatic congenital cytomegalovirus: disorders of language, learning, and hearing. Am. J. Dis. Child, 1982, 136:902-905.

2507. Wilson, J. Extensive vulval condylomata acuminata necessitating caesarean section. Aust. NZ. J. Obstet. Gynaecol., 1973, 13:121-124.

2508. Wilson, M.G. and Stein, A.M. Teratogenic effects of asian influenza. An extended study. J.A.M.A., 1969, 210:336-337.

2509. Wilson, T.E. Observations and comments on two outbreaks of abortion associated with IBR virus infection. Can. Vet. J., 1974, 15:227-229.

2510. Winters, W.D., Clouse, W.J. and Harris, S.C. Concomitant transfer of BCG-CW and canine virus antibodies from dams to pups. Dev. Comp. Immunol., 1981, 5:321-328.

2511. Withers, J. and Bradshaw, E. Preventing neonatal hepatitis-B infection. M.C.N., 1986, 11:270-272.

2512. Wittek, A.E., Yeager, A.S., Au, D.S. and Hensleigh, P.A. Asymptomatic shedding of herpes simplex virus from the cervix and lesion site during pregnancy. Correlation of antepartum shedding with shedding at delivery. Am. J. Dis. Child, 1984, 138:439-442.

2513. Witter, R.L. Natural mechanisms of controlling lymphotropic herpesvirus infection (Marek's disease) in the chicken. Cancer Res., 1976, 36:681-687.

2514. Witzleben, C.L. and Driscoll, S.G. Possible transplacental transmission of herpes simplex infection. Pediatrics, 1965, 36:192-199.

2515. Wohlgemuth, K., Leslie, P.F., Reed, D.E. and Smidt, D.K. Pseudorabies virus associated with abortion in swine. J. Am. Vet. Med. Assoc., 1978, 172:478-479.

2516. Wolf, A. and Cowen, D. Perinatal infections of the central nervous system. J. Neuropathol. Exp. Neurol., 1959, 18:191-243.

2517. Wolf, D.L. Herpes genitalis type 2 during pregnancy: report of case. J. Am. Osteopath. Assoc., 1977, 76:450-457.

2518. Wolff, S.M. The ocular manifestations of congenital rubella. A prospective study of 328 cases of congenital rubella. J. Pediatr. Ophthalmol., 1973, 10:101-141.

2519. Wong, F.C., Spearman, J.G., Smolenski, M.A. and Loewen, P.C. Equine parvovirus: initial isolation and partial characterization. Can. J. Comp. Med., 1985, 49:50-54.

2520. Wong, V.C., Ip, H.M., Reesink, H.W., Lelie, P.N., Reerink-Brongers, E.E., Yeung, C.Y. and Ma, H.K. Prevention of the HBsAg carrier state in newborn infants of mothers who are chronic carriers of HBsAg and HBeAg by administration of hepatitis-B vaccine and hepatitis-B immunoglobulin. Double-blind randomised placebo-controlled study. Lancet, 1984, 1:921-926.

2521. Wong, V.C., Lee, A.K. and Ip, H.M. Transmission of hepatitis B antigens from symptom free carrier mothers to the fetus and the infant. Br. J. Obstet. Gynaecol., 1980, 87:958-965.

2522. Woo, D., Cummins, M., Davies, P.A., Harvey, D.R., Hurley, R. and Waterson, A.P. Vertical transmission of hepatitis B surface antigen in carrier mothers in two west London hospitals. Arch. Dis. Child, 1979, 54:670-675.

2523. Woodman, C.B. and Weaver, J.B. Virological screening for herpes simplex virus (HSV) in late pregnancy. Lancet, 1986, 1:744.

2524. Woods, G.T., Hanson, L.E., Swenson, G.H. and Rhoades, H.E. Acute infectious bovine rhinotracheitis and salmonellosis in a dairy herd. J. Am. Vet. Med. Assoc., 1968, 152:1530-1535.

2525. Woods, G.T. and Mansfield, M.E. Transplacental migration of swine influenza virus in gilts exposed experimentally. Res. Commun. Chem. Pathol. Pharmacol., 1974, 7:629-632.

2526. Woods, G.T. and Mansfield, M.E. Experimental challenge of pregnant gilts with swine influenza virus after vaccination. Res. Commun. Chem. Pathol. Pharmacol., 1976, 15:787-795.

2527. Woody, N.C., Sistrunk, W.F. and Platou, R.V. Congenital syphilis: a laid ghost walks. J. Pediat., 1964, 64:63.

2528. Woolley, P.D., Monteiro, E., Wilson, J. and Kinghorn, G.R. Virological screening in late pregnancy in recurrent genital herpes. Lancet, 1986, 1:1336.

2529. Wrathall, A.E., Bailey, J., Done, J.T., Shaw, I.G., Winkler, C.E., Gibbons, D.F., Patterson, D.S. and Sweasey, D. Effects of experimental border disease infection in the pregnant sow. Zentralbl. Veterinarmed. [B], 1978, 25:62-69.

2530. Wrathall, A.E., Wells, D.E., Cartwright, S.F. and Frerichs, G.N. An inactivated, oil-emulsion vaccine for the prevention of porcine parvovirus-induced reproductive failure. Res. Vet. Sci., 1984, 36:136-143.

2531. Wright, E.P., Dyson, A.J. and Alaily, A. Infection with parvovirus during pregnancy. Br. Med. J. [Clin. Res.], 1985, 290:241.

2532. Wright, H.T. Jr. Congenital anomalies and viral infections in infants. The etiologic role of maternal viral infections. Calif. Med., 1966, 105:345-351.

2533. Wright, I. Congenital toxoplasmosis and deafness. An investigation. Pract. Otorhinolaryngol., 1971, 33:377-387.

2534. Wyatt, G.B. Pregnancy hepatitis in Libya. Lancet, 1977, 1:1204-1205.

2535. Wyatt, H.V. Poliomyelitis in the fetus and the newborn. A comment on the new understanding of the pathogenesis. Clin. Pediatr. (Phila.), 1979, 18:33-38.

2536. Wyks, H.W. The "rubella bulge" and vocational planning. Am. Ann. Deaf., 1980, 125:1022-1025.

2537. Wyll, S.A. Risk of rubella vaccination during pregnancy. Obstet. Gynecol., 1971, 38:641-642.

2538. Wyll, S.A. and Herrmann, K.L. Inadvertent rubella vaccination of pregnant women. Fetal risk in 215 cases. J.A.M.A., 1973, 225:1472-1473.

2539. Wynn-Williams, G. and Harrison, R.F. Infertility and rubella infection prophylaxis. Br. Med. J., 1974, 3:170.

# X

2540. Xu, Z.Y., Liu, C.B., Francis, D.P., Purcell, R.H., Gun, Z.L., Duan, S.C., Chen, R.J., Margolis, H.S., Huang, C.H. and Maynard, J.E. Prevention of perinatal acquisition of hepatitis B virus carriage using vaccine: preliminary report of a randomized, double-blind placebo-controlled and comparative trial. Pediatrics, 1985, 76:713-718.

# Y

2541. Yambao, T.J., Clark, D., Weiner, L. and Aubry, R.H. Isolation of cytomegalovirus from the amniotic fluid during the third trimester. Am. J. Obstet. Gynecol., 1981, 139:937-938.

2542. Yamouchi, T., St. Geme, J.W., Oh, W. and Davies, C.W.C. The biological and biochemical pathogenesis of mumps virus-induced embryonic growth retardation. Pediat. Res., 1975, 9:30.

2543. Yamouchi, T., Wilson, C. and St. Geme, J.W. Transmission of live attenuated mumps virus to the human placenta. N. Engl. J. Med., 1974, 290:710.

2544. Yang, S. and Halsted, C.C. Fever, seizures in a full-term newborn. Hosp. Pract. [Off.], 1985, 20:18, 20, 24.

2545. Yawn, D.H., Pyeatte, J.C., Joseph, J.M., Eichler, S.L. and Garcia-Bunuel, R. Transplacental transfer of influenza virus. J.A.M.A., 1971, 216:1022-1023.

2546. Yeager, A.S. Genital herpes simplex infections: effect of asymptomatic shedding and latency on management of infections in pregnant women and neonates. J. Invest. Dermatol., 1984, 83:53s-56s.

2547. Yeager, A.S., Arvin, A.M. and Sweeney, M.C. Absence of history of genital herpes simplex virus (HSV) infections in mothers of infected infants: source-asymptomatic shedding vs. primary infection. Pediatr. Res., 1982, 16:254A.

2548. Yeager, A.S., Davis, J.H., Ross, L.A. and Harvey, B. Measles immunization. Successes and failures. J.A.M.A., 1977, 237:347-351.

2549. Yeager, A.S., Grumet, F.C., Hafleigh, E.B., Arvin, A.M., Bradley, J.S. and Prober, C.G. Prevention of transfusion-acquired cytomegalovirus infections in newborn infants. J. Pediatr., 1981, 98:281-287.

2550. Yeager, A.S., Palumbo, P.E., Malachowski, N., Ariagno, R.L. and Stevenson, D.K. Sequelae of maternally derived cytomegalovirus infections in premature infants. J. Pediatr., 1983, 102:918-922.

2551. Yedloutschnig, R.J., Dardiri, A.H., Mebus, C.A. and Walker, J.S. Abortion in vaccinated sheep and cattle after challenge with Rift Valley fever virus. Vet. Rec., 1981, 109:383-384.

2552. Yip, D.M. and Baker, A.L. Liver diseases in pregnancy. Clin. Perinatol., 1985, 12:683-694.

2553. Ylinen, O. and Jarvinen, P.A. Parotitis during pregnancy. Acta Obstet. Gynecol., 1953, 32:121.

2554. Yohn, D.S., Olsen, R.G., Schaller, J.P., Hoover, E.A., Mathes, L.E., Heding, L. and Davis, G.W. Experimental oncornavirus vaccines in the cat. Cancer Res., 1976, 36:646-651.

2555. Young, A.B., Reid, D. and Grist, N.R. Is cytomegalovirus a serious hazard to female hospital staff? Lancet, 1983, 1:975-976.

2556. Young, A.W., Frankel, M. and Menegus, M. Papular (erosive) herpes genitalis. A clinical variant in sarcoid and pregnancy. Obstet. Gynecol., 1977, 50:730-732.

2557. Young, E.J. and Gomez, C.I. Enhancement of herpesvirus type 2 infection in pregnant mice. Proc. Soc. Exp. Biol. Med., 1979, 160:416-420.

2558. Young, E.J., Killam, A.P. and Greene, J.F. Jr. Disseminated herpesvirus infection. Association with primary genital herpes in pregnancy. J.A.M.A., 1976, 235:2731-2733.

2559. Young, R.L., Acosta, A.A. and Kaufman, R.H. The treatment of large condylomata acuminata complicating pregnancy. Obstet. Gynecol., 1973, 41:65-73.

2560. Yow, M.D. and Taber, L.H. Cytomegalovirus infections in the pregnant female and the newborn infant. Clin. Obstet. Gynecol., 1972, 15:993-1003.

2561. Yuill, T.M. The role of mammals in the maintenance and dissemination of La Crosse virus. Prog. Clin. Biol. Res., 1983, 123:77-87.

2562. Yukins, R.E. and Winter, F.C. Ocular disease in congenital toxoplasmosis in nonidentical twins. Amer. J. Ophthalmol., 1966, 62:44-46.

# Z

2563. Zakstelskaya, L.J., Arnaudova, V.I. and Yakhno, M.A. Humoral immunity factors to parainfluenza viruses type 1, 2 and 3 in infants under the age of one year. J. Hyg. Epidemiol. Microbiol. Immunol., 1969, 13:293-299.

2564. Zanetti, A.R., Ferroni, P., Tanzi, E. and Bergamini, F. Screening of pregnant women and hepatitis B prophylaxis in newborns. J. Virol. Methods, 1985, 10:341-347.

2565. Zanetti, A.R., Magliano, E.M., Tanzi, E., Ferroni, P., Pirovano, G., Pizzocolo, G., Pillan, N. and Zunin, C. HBIG immunoprophylaxis of babies born to HBsAg carrier mothers. Dev. Biol. Stand., 1983, 54:383-389.

2566. Zausmer, E. Congenital rubella. Pathogenesis of motor deficits. Pediatrics, 1971, 47:16.

2567. Zavell, P.M. Perinatal tips. Mich. Med., 1971, 70:1120.

2568. Zavoral, J.H., Ray, W.L., Kinnard, P.G. and Nahmias, A.J. Neonatal herpetic infection. A fatal consequence of penile herpes in a serviceman. J.A.M.A., 1970, 213:1492-1493.

2569. Zervoudakis, I.A., Silverman, F., Senterfit, L.B., Strongin, M.J., Read, S. and Cederqvist, L.L. Herpes simplex in the amniotic fluid of an unaffected fetus. Obstet. Gynecol., 1980, 55:16S-17S.

2570. Zhukovskij, A.N. The effect of influenza on the course of pregnancy and intra-uterine pathology of the foetus. Vopr. Virusol., 1972, 5:515.

2571. Zigas, V. Effect of kuru on pregnancy. Trop. Geogr. Med., 1973, 25:262-265.

2572. Zilliacus, H. and Pollanen, L. Sepsis, thrombotic purpura or hepatitis: A problem case. Ann. Chir. Gynaecol. Fenn., 1968, 57:398-401.

2573. Zinkham, W.H., Medearis, D.N. and Osborn, J.E. Blood and bone-marrow findings in congenital rubella. J. Pediatr., 1967, 71:512-524.

2574. Ziring, P.R., Fedun, B.A. and Cooper, L.Z. Thyrotoxicosis in congenital rubella. J. Pediatr., 1975, 87:1002.

2575. Ziring, P.R., Gallo, G. and Firegold, M. Chronic lymphocytic thyroiditis. Identification of rubella virus antigen in the thyroid of a child with congenital rubella. J. Pediatr., 1977, 90:419-420.

2576. Zuckerman, A.J. Maternal transmission of hepatitis. Nature (Lond.), 1974, 249:105.

2577. Zuckerman, A.J. The problem and control of hepatitis B infection in the fetus and the newborn. In: Krugman, S., Gershon, A.A., ed. Infections of the fetus and the newborn infant. New York, Liss, 1975, 97-113.

2578. Zuckerman, A.J. Perinatal transmission of hepatitis B. Arch. Dis. Child, 1984, 59:1007-1009.

2579. Zuelzer, W.W. Infantile toxoplasmosis, with a report of three cases, including two in which the patients were identical twins. Arch. Pathol., 1944, 38:1-19.

2580. Zuelzer, W.W. and Stulberg, C.S. Herpes simplex virus as the cause of fulminating visceral disease and hepatitis in infancy. Am. J. Dis. Child, 1952, 83:421-439.

# Subject Index

## About the Compiler

ERNEST L. ABEL is Director of Operations and Professor of Obstetrics at the C.S. Mott Center for Human Growth and Development, Wayne State University. He is the author of *A Dictionary of Drug Abuse Terms and Terminology* (Greenwood Press, 1984), and the compiler of numerous bibliographies, including *Homicide, Behavioral Teratology, Fetal Alcohol Exposure and Effects,* and *Narcotics and Reproduction,* all published by Greewood Press.

www.ingramcontent.com/pod-product-compliance
Ingram Content Group UK Ltd.
Pitfield, Milton Keynes, MK11 3LW, UK
UKHW020239290926
15470UKWH00002B/624

9 780313 264399